Model-Based Control Engineering - Recent Design and Implementations for Varied Applications

Edited by Umar Zakir Abdul Hamid and Ahmad `Athif Mohd Faudzi

Published in London, United Kingdom

IntechOpen

Supporting open minds since 2005

Model-Based Control Engineering - Recent Design and Implementations for Varied Applications
http://dx.doi.org/10.5772/intechopen.94801
Edited by Umar Zakir Abdul Hamid and Ahmad `Athif Mohd Faudzi

Contributors
Ismit Mado, Derrick Rollins, Sharaf Eddine Kramti, Jaouher Ben Ali, Hugo Andre, Eric Brhhoefer, Mounir Sayadi, Murat Sahin, Umar Zakir Abdul Hamid, Ahmad 'Athif Mohd Faudzi

First published in London, United Kingdom, 2022 by IntechOpen
IntechOpen is the global imprint of INTECHOPEN LIMITED, registered in England and Wales, registration number: 11086078, 5 Princes Gate Court, London, SW7 2QJ, United Kingdom
Printed in Croatia

British Library Cataloguing-in-Publication Data
A catalogue record for this book is available from the British Library

Additional hard and PDF copies can be obtained from orders@intechopen.com

Model-Based Control Engineering - Recent Design and Implementations for Varied Applications
Edited by Umar Zakir Abdul Hamid and Ahmad `Athif Mohd Faudzi
p. cm.
Print ISBN 978-1-83969-590-2
Online ISBN 978-1-83969-591-9
eBook (PDF) ISBN 978-1-83969-592-6

For EU product safety concerns:
IN TECH d.o.o., Prolaz Marije Krucifikse Kozulić 3, 51000 Rijeka, Croatia, info@intechopen.com or visit our website at intechopen.com.

Meet the editors

Umar Zakir Abdul Hamid, Ph.D., has been working in the future mobility (connected and autonomous vehicle) field since 2014 with various teams in different countries, including Malaysia, Singapore, Japan, Finland, and Sweden. He has led, managed, and worked on more than thirty driverless technology, robotics, and automotive projects. Previously, he led a team of twelve engineers working in the Autonomous Vehicle Software Product Development with Sensible 4, Finland. Dr. Hamid is one of the recipients of the Finnish Engineering Award 2020 for his contributions to the development of all-weather autonomous driving solutions. He is an aspiring automotive thought leader and is often invited as a guest and keynote speaker at industrial and technical events. With more than thirty scientific publications as author and editor to his credit, Dr. Hamid actively participates in global automotive standardization efforts where he is a secretary for a Society Automotive Engineers (SAE) committee. Since the end of Summer 2021, he has been working as the Lead of Strategic Planning for CEVT AB in Sweden.

Prof. Ir. Ts. Dr. Ahmad `Athif Mohd Faudzi received a BEng in Computer Engineering and an MEng in Mechatronics and Automatic Control from Universiti Teknologi Malaysia, and a Dr.Eng. in System Integration from Okayama University, Japan, in 2004, 2006, and 2010, respectively. He was a visiting research fellow at the Tokyo Institute of Technology from 2015 to 2017. He was also a fellow in the Academia-Industry Talent Exchange Programme (AI-xChange): Ceo@Faculty Programme 2.0 'Coached by the Pro' under Todd Ashton, CEO of Ericsson Malaysia Sdn. Bhd. from 2018 to 2019. He is currently the director of the Centre for Artificial Intelligence and Robotics (CAIRO), Universiti Teknologi Malaysia. He is a Professional Engineer with Practicing Certificate (PEPC), a Charted Engineer (CEng), president for Persatuan Saintis Muslim Malaysia (PERINTIS), committee member of the Malaysia IEEE Robotics and Automation Society (RAS), and a member of ASM-YSN, Akademi Sains Malaysia Special Interest Group ASM-SIG Biodiversity, and ASM-SIG Robotics. He is the leader of the R&D subgroup Malaysia National Robotic Roadmap. In 2020, he received the Top Research Scientist Malaysia (TRSM) award in Robotics. He is mainly engaged in the research fields of actuators (pneumatic, soft mechanism, hydraulic, and motorized actuators) with a focus on robotics, bioinspired robotics, and biomedical applications.

Contents

Preface

Progress in industrialization and automation engineering is creating many new opportunities in the autonomous systems industry. Thus, there is a demand for control engineering across numerous industries. For example, in the automotive industry, "software-defined" vehicles, which are vehicles whose features and functions are primarily enabled through software, are being researched and developed. In the maritime world, companies like Wärtsila and ABB are working on autonomous vessel technologies. Similar progress can be seen in aircraft technologies with the innovation of flying taxis.

With the uncertain and highly nonlinear dynamics of the real world where these new technologies will be deployed, a reliable control strategy is necessary. One of these methods is model-based control engineering. This book provides a high-level discussion on model-based control engineering and its various applications.

The book is divided into three sections. The first section includes an introductory chapter. The second section, "Identification and Prediction," discusses topics such as parameter identification and particle filter-based approaches for prediction purposes. The final section, "Modelling and Optimization," contains several discussions on model predictive control as well as control optimization.

The editors are experienced researchers in the control engineering field, currently based in Sweden and Malaysia. We would like to thank all the reviewers who contributed to the creation of this book: Erkan Adalı, Dr. Ayush Jain, Zejiang Wang, Dr. Vinay Pandey, and Dr. Mathias Metzler.

Dr. Umar Zakir Abdul Hamid
China Euro Vehicle Technology AB,
Gothenburg, Sweden

Dr. Ahmad 'Athif Mohd Faudzi
Professor,
Universiti Teknologi Malaysia,
Johor Bahru, Malaysia

Section 1

Model-Based Control Engineering

Chapter 1

Introductory Chapter: Model-Based Control Engineering and Its Significance for Automation Technology and Autonomous Systems

Umar Zakir Abdul Hamid and Ahmad 'Athif Mohd Faudzi

1. Introduction

The progress in industrialisation and automation engineering creates a lot of new opportunities in the autonomous systems industry [1]. For example, in the automotive world, the term 'software-defined' vehicle is starting to accumulate more discussions [2]. It refers to the future where the vehicle value will be defined more on its software, in addition to the hardware platform.

Across the industries, we are seeing the need for control engineering knowledge continuously in demand. For example, in the maritime world, companies such as Wärtsila and ABB are working on autonomous vessels technologies [3, 4]. Furthermore, similar progress can be seen in aircraft technologies with the advances of flying taxis innovations [5].

For all of these technologies to be delivered safely to the public, it needs to be able to be operating continuously while being exposed to the uncertainties of the systems. Therefore, one of the methods to address the mentioned issues is model-based control engineering adoption to yield a reliable performance of the systems.

2. Model-based control engineering

Model-based control engineering facilitates the development of complex systems using a model-based design approach. Among the notable examples of application in the control engineering is model predictive control (MPC), where it is usually adopted to address the complex behaviour of nonlinear dynamic systems (**Figure 1**). MPC is a highly studied topic in the control system field. The inclusion of the dynamic models of a plant or process into the formulation aids in improving the control system performance.

In typical model-based control engineering works, the plants can be modelled *via* several means. Once the obtained model has been verified, the controller and algorithm development can be driven with it. These include the validation and verifications of the algorithms with the modelled plant. For example, a vehicle collision avoidance control system development can be simulated with a reliable vehicle model [9, 10]. With the advances of computing devices, a lot of researchers

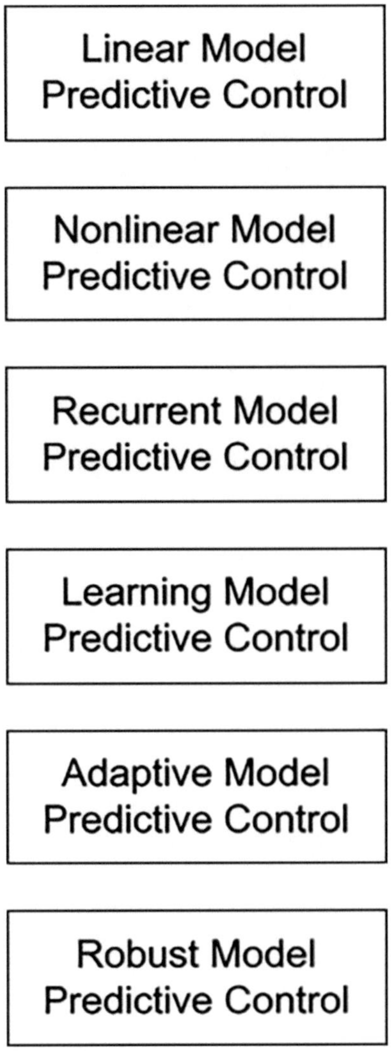

Figure 1.
Different types of model predictive control strategies [6–8].

are now integrating model-based applications with machine learning applications for complex autonomous systems [11, 12].

With model-based control engineering, a lot of benefits can be attained. For example, the time-to-market of the product can be reduced by solving the bottleneck subjects between hardware readiness and control system development [13, 14]. Furthermore, a model with good fidelity will allow for simulation-based testing of the control systems in different scenarios during the system testing stage. Despite this, model-based control engineering encounters more challenges too. Among the topics that are related to model-based control engineering are system identifications, modelling, and optimization.

3. Aim of this book

With advances in computational devices, model-based control applications, particularly MPC has started to gain recognition by practitioners in varied industrial sectors such as chemical engineering and industrial plant applications, and recently in robotics and autonomous vehicles, among many others. However, despite the advance and progress, implementation in uncertain environments and highly nonlinear scenario remains challenging.

This book aims to provide model-based control engineering topics high-level discussions to the generic audience with varied use cases. It is hoped to provide a good overview of model-based control engineering for interested readers. As we are seeing more autonomous systems entering the market, the editors of this book believe the discussions made in this book will be useful for the readers' knowledge of model-based control engineering.

As this book is aimed to be brief and cover different perspectives, the editors are also encouraging interested readers to read more extensive discussions on the theoretical part of model-based control engineering in these works [15–17].

Acknowledgements

The editors would like to thank all the reviewers and authors involved in the works.

Author details

Umar Zakir Abdul Hamid[1*] and Ahmad 'Athif Mohd Faudzi[2]

1 China Euro Vehicle Technology AB, Sweden

2 Universiti Teknologi Malaysia, Malaysia

*Address all correspondence to: editorumarzakir@gmail.com

References

[1] Shahrdar, Shervin, Luiza Menezes, and Mehrdad Nojoumian. "A Survey on Trust in Autonomous Systems". In: *Science and Information Conference*. Cham:Springer; 2018. pp. 368-386

[2] Shapiro D. The future of automotive is software-defined. ATZelectronics worldwide. 2020;**15**(12):72-72

[3] Quraishi H, Huntly-Playle I, Chung C, Pedersen T. Approach to autonomous technology development at Wärtsilä. Marine Engineering. 2020;**55**(6):677-684

[4] Elings, Adriaan, Burger Arie Wolst, Joris de Man, and Koen Petersen. "Autonomous sailing interconnectivity." 2016. Available from: http://maritimesymposium-rotterdam.nl/uploads/Route/Autonomous%20sailing%20INTERCONNECTIVITY.pdf

[5] Lenton D. The measure of Volocopter flying taxi. Engineering & Technology. 2018;**13**(7/8):10-11

[6] Holkar KS, Waghmare LM. An overview of model predictive control. International Journal of Control and Automation. 2010;**3**(4):47-63

[7] Liu, Zhengyu, Jingliang Duan, Wenxuan Wang, Shengbo Eben Li, Yuming Yin, Ziyu Lin, Qi Sun, and Bo Cheng. "Recurrent Model Predictive Control." arXiv preprint arXiv:2102.10289. 2021. Available from: https://arxiv.org/abs/2102.10289

[8] Lenz I, Knepper RA, Saxena A. DeepMPC: Learning deep latent features for model predictive control. Robotics: Science and Systems. July 2015;**10**:312. Available from: http://www.roboticsproceedings.org/rss11/p12.pdf

[9] Tram, Tommy, Ivo Batkovic, Mohammad Ali, and Jonas Sjöberg. "Learning when to drive in intersections by combining reinforcement learning and model predictive control." In: 2019 IEEE Intelligent Transportation Systems Conference (ITSC), IEEE. 2019 pp. 3263-3268. Available from: https://ieeexplore.ieee.org/abstract/document/8916922

[10] Lyu Y, Jinwen H, Chen BM, Zhao C, Pan Q. Multivehicle flocking with collision avoidance via distributed model predictive control. IEEE transactions on cybernetics. 2019;**51**(5):2651-2662

[11] Mehndiratta, Mohit, Efe Camci, and Erdal Kayacan. "Automated tuning of nonlinear model predictive controller by reinforcement learning." In: *2018 IEEE/RSJ International Conference on Intelligent Robots and Systems (IROS)*, IEEE. 2018. pp. 3016-3021. Available from: https://ieeexplore.ieee.org/abstract/document/8594350

[12] Bujarbaruah, Monimoy, Xiaojing Zhang, Ugo Rosolia, and Francesco Borrelli. "Adaptive MPC for iterative tasks." In: 2018 IEEE Conference on Decision and Control (CDC), IEEE. 2018. pp. 6322-6327. Available from: https://ieeexplore.ieee.org/abstract/document/8618694

[13] Kelemenová T, Michal Kelemen L, Miková VM, Prada E. Tomáš Lipták, and Frantisek Menda. "model based design and HIL simulations." American journal of. Mechanical Engineering. 2013;**1**(7):276-281

[14] Butzin, Björn, Frank Golatowski, Christoph Niedermeier, Norbert Vicari, and Egon Wuchner. "A model based development approach for building automation systems." In: Proceedings of the 2014 IEEE Emerging Technology and Factory Automation (ETFA), IEEE. 2014. pp. 1-6. Available from: https://ieeexplore.ieee.org/abstract/document/7005365

[15] Jarzębowska E. Model-Based Tracking Control of Nonlinear Systems. Boca Raton: CRC Press; 2012

[16] Model-Based Control: Bridging Rigorous Theory and Advanced Technology edited by Paul M.J. van den Hof, Carsten Scherer. Available from: https://link.springer.com/book/10.1007/978-1-4419-0895-7

[17] Agachi, Paul Serban, Zoltán K. Nagy, Mircea Vasile Cristea, and Árpád Imre-Lucaci. Model Based Control: Case Studies in Process Engineering. John Wiley & Sons. 2007. Available from: https://biust.pure.elsevier.com/en/publications/model-based-control-case-studies-in-process-engineering

Section 2

Identification and Prediction

Chapter 2

Identification of Predicted Load Cluster Pattern Power Generation Parameters Based on Descriptive Time Series Analysis

Ismit Mado

Abstract

This chapter describes the process of identifying a power generation system. This is important because in principle the system parameters as a whole are not linear and uncertain. For this reason, it is necessary to carry out an identification process using an experimental approach that is able to represent the system as a whole. The technique used in this identification process is Prediction Error Minimization (PEM) as a tool available in Matlab. Identification is done by simulating changes in the value of frequency, voltage and electrical power due to changes in load. The change in load over time is a characteristic of the time series pattern. Through descriptive analytic approach, the cluster load is patterned for each load operating condition. Through load clusters, the identification results of power generation systems are obtained based on their operating conditions. This chapter presents validated parameter estimates for each change in instantaneous load conditions. The simulation results obtained better performance between the actual output and the identification model, namely the calculation of the Intergal Absolute Error (IAE), with MAPE for the average frequency value of 73.95 percent, nominal voltage of 0.23 percent, and electric power of 23.46 percent.

Keywords: Identification parameters, Validation, Descriptive analysis, Clustering model, Electric loads, Power generation system

1. Introduction

In reality, existing systems are a combination of linear and nonlinear models. Modeling with mathematical derivation is done with a lot of neglect of unmeasured parameters. The calculation is carried out around the linear area only, while the nonlinear area is not much heeded or neglected. Therefore, the results of the model derived based on the laws of physics or mathematical derivation, are still not very effective to be applied directly in the field.

Apart from the method of deriving the laws of physics or deriving it mathematically, there are also other methods for modeling, namely by using the identification method. This method will model a system as a whole, both linear and nonlinear parameters. All of these are considered to be one integrated system. Identification is an approach process by means of mathematical modeling of an unknown system through learning of data collected from previous experiments as well as input–output data from pre-existing controlled systems. The results obtained are parameter values in the form of a mathematical model. The data is compared with the actual model, wherein the difference between the two systems is used as an objective function which will later be minimized. In principle, the identification process is an introduction to a system to be controlled. Identification is one of the important factors that support success in engineering a control system that is stable, robust and able to represent the actual model in the field.

Matlab has a very effective identification tool for simulations. Several studies on parameter identification have been carried out through one of them the System Identification Toolbox (SIT) package. Initial research which is the reference for this writing as has been done by Fruk et al. [1]. The model can be determined by adjusting the assumed process model settings, until the modeling is satisfactory and accurate according to the input data in the SIT package. Research has also been carried out to identify the dynamic model parameters of a permanent magnet direct current (PMDC) DC motor [2]. The process of identifying DC motor parameters in the last year has also been carried out and published. This study identifies DC motor parameters whose initial values are estimated by MATLAB/ Simulink based on a Genetic Algorithm [3]. Identification research on large-scale power generation systems has been carried out to test the frequency value of the load shedding test on the 39 bus hydraulic turbogenerator system [4]. This identification process is carried out in the form of a Matlab simulation.

The identification in this paper is a variant of the single machine infinite bus (SMIB) parameters due to load changes. The fluctuation of electric power at the load center is very important to analyze so that the stability of the power generation system can be maintained optimally. Changes in load over time can be identified as minor disturbances in the generating system. Minor disturbances or so called small-signal stability studies occur in the operating conditions of the system after the first swing at which time the response of control equipment such as governors, AVR, and auxiliary devices has been taken into account [5, 6]. Load fluctuation results in changes in the value of the frequency and voltage in the power generation system so that the stability of the system will be disturbed.

The load characteristics of the period of use whether it is used by household, commercial, industrial and public loads are necessary so that fluctuations in the loading system can be analyzed. Changes in electric load form a continuous pattern and are time series in nature. Through a descriptive analytical statistical approach, the load pattern is analyzed so that the operating conditions of the power generation system are obtained in the form of a continuous load cluster pattern. To achieve the research objectives, steps were taken to identify any changes in electrical power at the load center. Changes in load are classified through the identification of SMIB parameters using the reduced-order model approach.

2. Research methods

This research aims to identify and validate the power generation system parameters. The identification process uses the prediction error minimization approach to obtain a more accurate mathematical model.

In this research, identification is carried out for all dynamic conditions due to changes in electric power on the load side. Namely, creating a load cluster pattern based on the analytical descriptive method in a time series approach.

The achievement that will be produced in this research is the identification model equation for each cluster in the form of a state space matrix equation. So that the results of this research will be useful in designing controls on a power generation system that is more efficient, optimal response and robust.

2.1 Proposed research

Mathematical models through physical analysis will not provide accurate identification results, because there are parameters that can only be obtained through an experimental process. So it is necessary to approach the system model. The system model approach can be carried out through the identification and validation process stages. System identification is defined as a method used to obtain an approach model from the actual system through evaluation of input–output measurement data [7]. In other words, the process of identifying a system is a combination of two efforts, namely the effort to form a mathematical model and estimate the optimal parameter value through experimental steps. System identification in general can be described as in **Figure 1** below.

According to Law and Kelton, validation is the process of determining whether the simulation conceptual model is really an accurate representation of the real system being modeled [8]. Model validation can also be said as a step to test whether the identified model can represent the real system correctly. A model can be said to be valid when it does not have a significant difference with the real system which is observed either from its characteristics or from its behavior.

It is important to identify and validate the power generation system in determining the load change pattern. The dynamic load characteristics will be patterned in the form of clusters based on statistical analysis. Load every moment is a time series dynamic behavior. A time series is a series of observations carried out sequentially based on time [9]. The observation process is carried out at the same intervals, for example in hourly, daily, weekly, monthly, yearly intervals, or other intervals. There are two objectives of time series analysis, namely to model the stochastic mechanisms contained in the observations and to predict the value of future observations.

Electric load data can be viewed as a reality of stochastic processes [9]. Where statistical phenomena are arranged in time order based on the law of probability. If the time series observation is denoted by Z_t, where $t \in A$ with A the set of natural numbers. According to Wei, the stochastic process is a time-based data group composed of random variables $Z(\omega, t)$ where ω is the sample space and t is the time index [10]. The distribution functions of the random variables $Z_{t_1}, Z_{t_2}, \dots, Z_{t_n}$ are as follows:

$$F\left(z_{t_1}, z_{t_2}, \dots, z_{t_n}\right) = p\{\, \omega : z\left(\omega, t\right) \le z_{t_1}, \dots, z\left(\omega, t_n\right) \le z_{t_n}\} \quad (1)$$

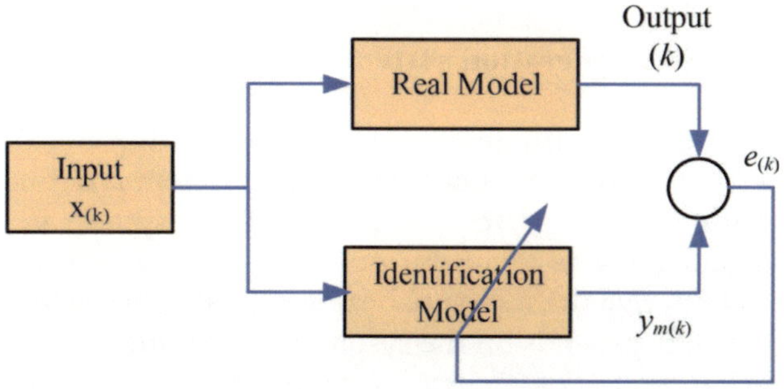

Figure 1.
Block diagram of the system identification process.

Observation of $Z_1, Z_2, \dots, Z_n$ is a stochastic process, so the random variables $Z_{t_1}, Z_{t_2}, \dots, Z_{t_n}$ are said to be stationary in the distribution if:

$$F\left(z_{t_1}, z_{t_2}, \dots, z_{t_n}\right) = F\left(z_{t1+k}, z_{t2+k}, \dots, z_{tn+k}\right) \quad (2)$$

A model like the one above is called a stochastic process, because of the sequential observations that are arranged in time. The variant of the electric load due to use in consumers can be viewed as an approach to the electrical load cluster pattern. The main purpose of this cluster pattern is to determine the operating conditions of the power generation system. Analytical descriptive method is a statistical approach that can describe a continuous load state. This method is concerned with collecting and presenting a data set so that it provides useful information [11]. Analysis of this cluster data processing into information containing a set of electrical load characteristics that can be summed up numerically.

This descriptive analysis includes several things, namely: first, frequency distribution, namely the arrangement of data according to certain categories in a systematically arranged list. Second, the measurement of central tension, which is a statistical analysis that specifically describes a representative score, includes data frequency figures such as mode, mean, median or arithmetic mean. Third, the measurement of variability, namely the degree of spread of variable values from a central tendency in a distribution. Variability is also known as dispersion. Variability can be measured through measurements: range, mean deviation, and standard deviation [12].

Cluster pattern groups in the same interval will make it easier to analyze load changes and be able to provide intervals of variants of the distribution of electric power from the power plant to the load. This analysis is important in achieving the planning goals and schedules of a more efficient and optimal generating system in maintaining the balance of the electric power system.

2.2 Physical model of the power generation system

The physical model used in this study is the SMIB model. The SMIB model refers to Park modeling, with the following criteria: negligible stator resistance, balanced system conditions and negligible core saturation of the generator, and the load is considered a static load [13]. This model refers to the synchronous machine that was introduced by De Mello and Concordia [14]. Recent studies still refer to this single engine model as has been done by [15–17].

The electric power produced by the power plant must be balanced with the electric power absorbed at the load center. By applying a variant of the electric load pattern and the load cluster approach, this research is able to maintain a balance between the electrical power supplied and the electric power consumed at the load center.

2.3 Identification power generation system

Parameter identification through derivation of the mathematical equation of the SMIB model. The mathematical model of a dynamic system is defined as a set of mathematical equations that represent the dynamics of a system accurately or at least close to the characteristics of the dynamic system. As a first step in analyzing a dynamic system is to derive the mathematical model. The mathematical model of each system will vary, it depends on the system components. Previous studies have derived mathematical models of generating systems to support the design of small signal stability control [18].

The identification approach uses a system identification toolbox with prediction error minimization (PEM) black-box identification techniques. PEM builds a measured input–output system mathematical model with the aim of updating the initial model as a reference model.

The PEM function also handles multiple-input-single-output structures in the form of polynomial representation of transfer functions

$$A(q)y(t) = \frac{B_1(q)}{F_1(q)}u_1(t - n\,k_1) + \ \dots\ + \frac{B_{nu}(q)}{F_{nu}(q)}u_{nu}(t - n\,k_{nu}) + \frac{C(q)}{D(q)}\,e(t) \qquad (3)$$

Where A, B, F,C and D are polynomials in the operator delay. Here, the numbers n_a and n_b are the orders of the respective polynomials. The number nk is the number of delays from input to output. With n_y output channels and nu input channels. An Output-Error structure is obtained as$e(t)$. In this study, using discrete time with a 3rd order approach.

3. Results of the parameter identification

To reach the results stage of this research, the steps taken in this study are simulations of the identification of the power generation system parameters, the results of time series analysis, validation, and simulation of the validation parameters compared to the actual model.

3.1 Parameter modeling

Based on **Figure 1**, the mathematical model of the SMIB system is represented in the form of the following state space Eq. [14]:

$$\dot{x} = A\,x + B\,u \text{ and } y = C\,x \qquad (4)$$

Where, x is the state variable, $n \times 1$, u is the input variable, $m \times 1$, y is the output variable, $r \times 1$, A is the system matrix, $n \times n$, B is the control matrix, $n \times m$, and C is the measurement matrix, $n \times m$.

The state space equation, where the state variable x is defined as

$$x = \begin{bmatrix} \Delta Y & \Delta T_m & \Delta\delta & \Delta\omega & \Delta E'_q & \Delta v_F \end{bmatrix}^T \qquad (5)$$

and the output variable y as

$$y = \begin{bmatrix} \Delta Y & \Delta T_m & \Delta P & \Delta\omega & \Delta v & \Delta v_F \end{bmatrix}^T \qquad (6)$$

where,

$$\text{Matrix } A_{ii} = \begin{bmatrix} -\dfrac{1}{T_{gu}} & 0 & 0 & -\dfrac{K_{gu}}{T_{gu}\,R} & 0 & 0 \\ \dfrac{1}{T_{tu}} & -\dfrac{1}{T_{tu}} & 0 & 0 & 0 & 0 \\ 0 & 0 & 0 & \omega_o & 0 & 0 \\ 0 & 0 & -\dfrac{K_1}{M} & -\dfrac{D}{M} & -\dfrac{K_2}{M} & 0 \\ 0 & 0 & -\dfrac{K_4}{T'_{do}} & 0 & -\dfrac{1}{K_3\,T'_{do}} & \dfrac{1}{T'_{do}} \\ 0 & 0 & -\dfrac{K_A K_5}{T_A} & 0 & -\dfrac{K_A K_6}{T_A} & -\dfrac{1}{T_A} \end{bmatrix}$$

$$\text{Matrix } B_{ii} = \begin{bmatrix} \frac{K_{gu}}{T_{gu}} & 0 \\ 0 & 0 \\ 0 & 0 \\ 0 & 0 \\ 0 & 0 \\ 0 & \frac{K_A}{T_A} \end{bmatrix} \text{ and Matrix } C_{ii} = \begin{bmatrix} 1 & 0 & 0 & 0 & 0 & 0 \\ 0 & 1 & 0 & 0 & 0 & 0 \\ 0 & 0 & K_1 & D & K_2 & 0 \\ 0 & 0 & 0 & 1 & 0 & 0 \\ 0 & 0 & K_5 & 0 & K_6 & 0 \\ 0 & 0 & 0 & 0 & 0 & 1. \end{bmatrix}$$

And for the output state variable of the equation that satisfies:

$$\Delta P = K_1\Delta\delta + K_2\Delta E_q' + D\Delta\omega \tag{7}$$

$$\Delta v = K_5\Delta\delta + K_6\Delta E_q' \tag{8}$$

Where, ΔY is the change in turbi valve height, ΔT_m is the change in mechanical torque, $\Delta\delta$ is the change in rotor angle, $\Delta\omega$ is the change in rotor angular velocity, $\Delta E_q'$ is the change in generator transient voltage, Δv_F is the change in excitation output voltage, ΔP is the change in generator electrical power, and Δv is the change in terminal voltage.

3.2 Time series analysis based load cluster

The dynamics of the electrical load is a series of data calculated every half hour on the SMIB that transmits power to the load center. The operating points were selected based on load cluster modeling using descriptive analytical statistical methods. The cluster interval range is simply implemented between the minimum value of electrical load, quartile 1, middle value, quartile 3, and maximum value as the operating point under load conditions. To model the cluster pattern based on the distribution of electrical load data as shown in **Figure 2**, this study uses the Minitab software.

Load variation is defined as a disturbance mechanism that occurs in the system due to changes in electrical power at the load center represented by changes in load groups. While the input from the turbine side and constant excitation. This modeling is represented by the input signal (u) which is on the turbine side (Δu_{gu}) and the excitation side (Δu_E) with a reference a signal of 0.5 sin (0.1 t) + 0.5 pu. Output (y) which represents the signal change in frequency value ($\Delta\omega$), change in terminal voltage on the generator bus, (Δv) and generator electrical power (ΔP) in the form of 1 pu.

The input change load is in the form of a cluster pattern which represents the cluster load model in the form of a sinusoidal signal equation. As shown in **Figure 3**, the cluster pattern in the form of a variation in the load model is set manually. The cluster charge signal is a reference input with a signal pattern: cluster 1 represents a 0.125 sin (0.5 t) +0.125 signal, cluster 2 is 0.125 sin (0.5 t) +0.375, cluster 3 is 0.125 sin (0.5 t) +0.625, and cluster 4 of 0.125 sin (0.5 t) +0.875 in 1 pu.

In terms of the discrete state model equation, the identification results are expressed by:

$$x(kT+1) = aM1\, x(kT) + bM1\, u(kT) \tag{9}$$

$$y(kT) = cM1\, x(kT) \tag{10}$$

For all identification processes using a sampling interval of 0.1 s with a discrete equation mode of order 3. The program listing on the system identification toolbox is as follows

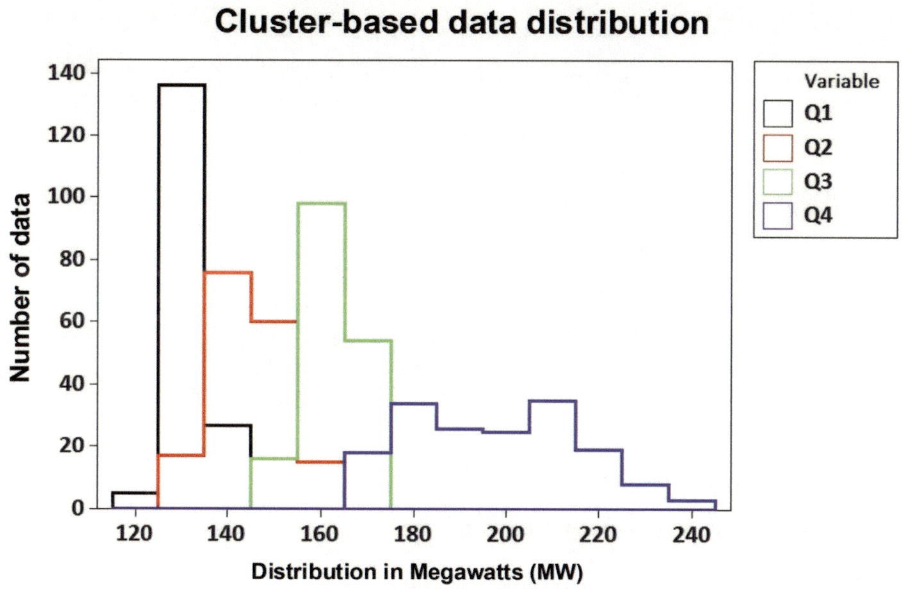

(a)

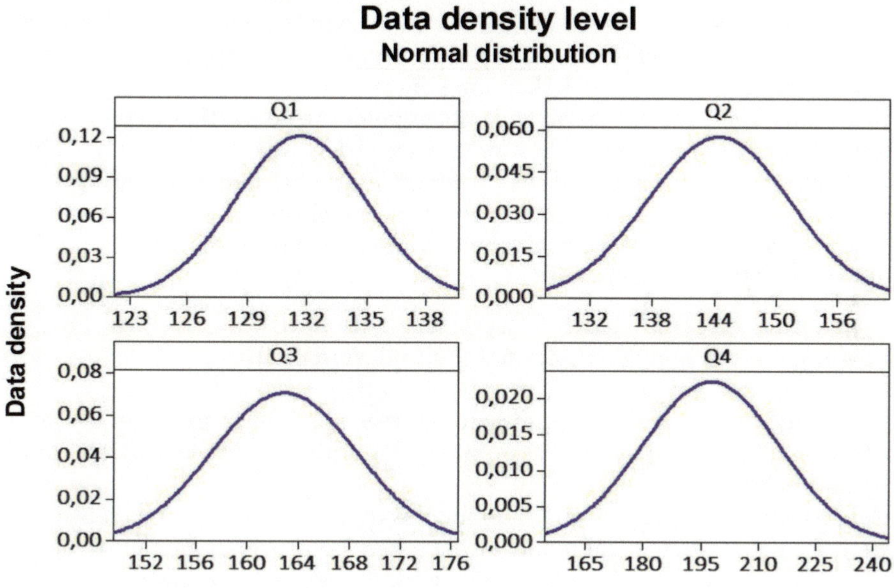

(b)

Figure 2.
(a) The distribution pattern of the cluster model data, (b) the level of data density.

```
S%Identification ModelSMIB
dat = idddata(y11 y12 y13],u11 u12],0.1);
SMIB = pem(dat,nx, ...
      'DisturbanceModel','none', ...
      'InitialState','zero');
[aM1d, bM1d, cM1d, dM1d, ke] = ssdata(SMIBDskrt);
aM1d
bM1d
cM1d
```

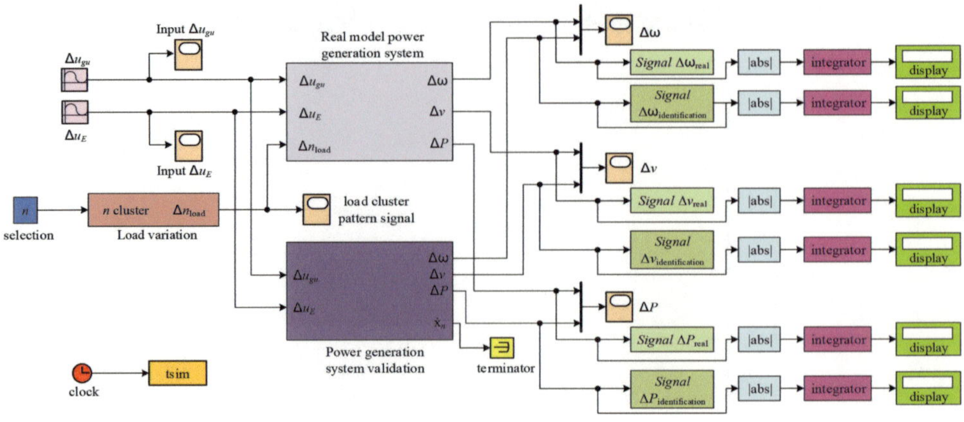

Figure 3.
Block diagram of the SMIB validation.

3.3 Validation power generation system

Power generation system validation model represents the mathematical model of the actual model and the 3rd order approach model. The representation of the simulation model approach with Matlab/Simulink is in the following block diagram.

The sub-system model of the generation system validation is shown in the following Matlab/Simulink block diagram (**Figure 4**).

The results of the identification and validation of the generating system are expressed in the form of the equation state space order 3 matrices A, B, and C below.

Modeling the identification of the power generation system parameters through the simulation of loading of electrical power in each cluster obtained a linear model of the input state matrix (A matrix), the output state matrix (C matrix) and the control signal state matrix (B matrix).

3.4 Power generation system parameter simulation results

The identification result is a state space matrix equation as shown in Eqs. 9 and 10 in **Table 1** above. The identification process is outlined in a flowchart as shown in **Figure 5** below:

The identified state space equation model is then validated as a reference equation for the mathematical model of the power generation system load cluster parameters compared to the actual model, with the following results:

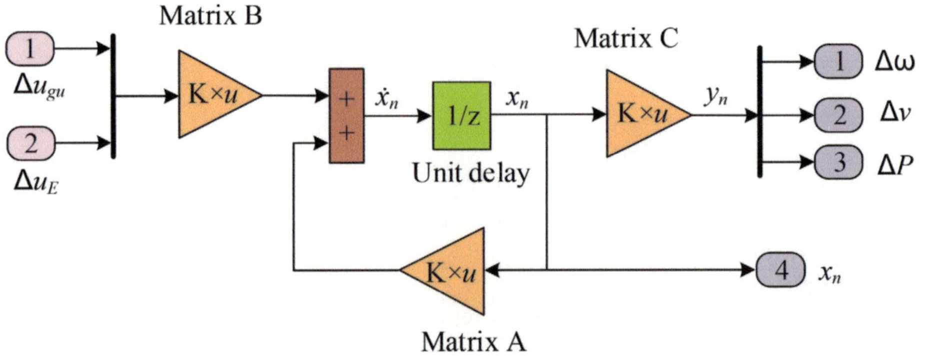

Figure 4.
Sub-system model of generation system validation.

DOI: http://dx.doi.org/10.5772/intechopen.99126

Cluster	Matrix A			Matrix B		Matrix C		
1	1, 0006	0, 4699	−0, 0567	−1, 1014	−0, 4644	−0, 0333	−0, 0208	0, 0594
	−0, 0562	0.8521	0, 0317	0, 0906	−0, 8310	−0, 0179	−0, 1015	−0, 0531
	−0, 0208	0, 1863	0, 9662	−0, 5709	−0, 7381	0, 2325	0, 5239	−0, 4321
2	0, 9499	0, 6371	−0, 0459	−0, 6493	−0, 4438	−0, 0056	0, 0043	−0, 0084
	−0, 0397	0, 8488	−0, 0070	0, 0823	−0, 9872	−0, 0246	−0, 1114	0, 0185
	0, 0234	−0, 3796	0, 9921	0, 4765	0, 6902	0, 1572	0, 3463	0, 2393
3	1, 0513	0, 5022	0, 2361	0, 1099	−0, 9611	0, 0022	0, 0125	0, 0049
	−0, 0662	0, 8710	−0, 0124	−0, 2356	−0, 6486	−0, 0735	−0, 1159	−0, 0107
	0, 0251	−0, 1192	0, 8908	0, 5362	0, 5751	0, 0960	0, 4697	0, 2131
4	0, 8189	−1, 1845	−0, 5249	0, 1141	2, 4466	−0, 0188	−0, 1593	−0, 1006
	0, 0910	1, 3565	0, 2761	−0, 1583	0, 7748	0, 0151	0, 0293	−0, 068
	−0, 1205	−0, 4570	0, 5857	0, 2317	−1, 7243	0, 0310	−0, 2124	−0, 1050

Table 1.
Cluster based state space matrix.

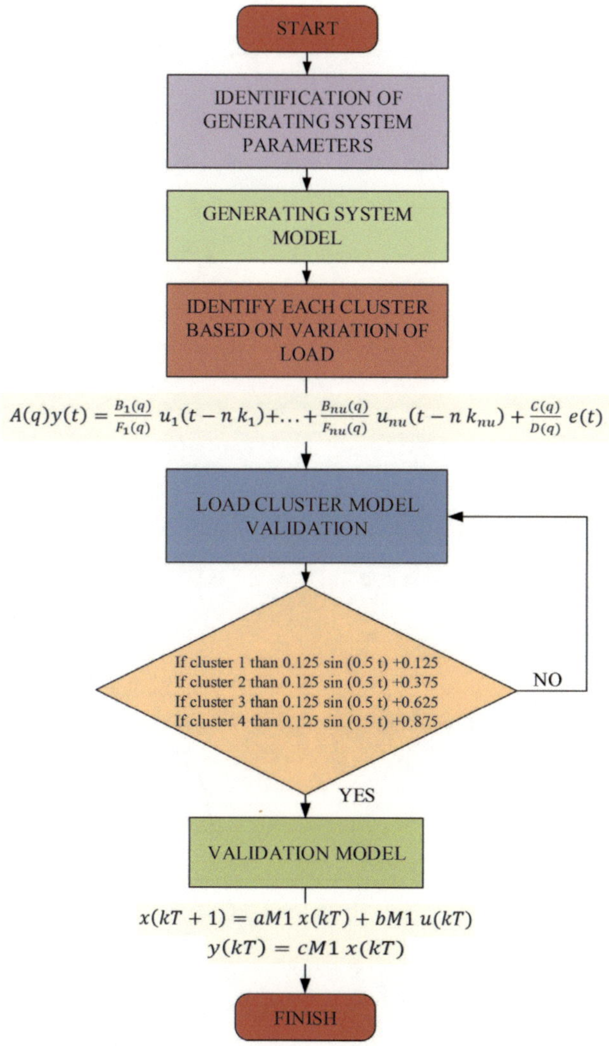

Figure 5.
Flowchart research.

1. Validate the following cluster 1 power generation system parameters (**Figure 6**):

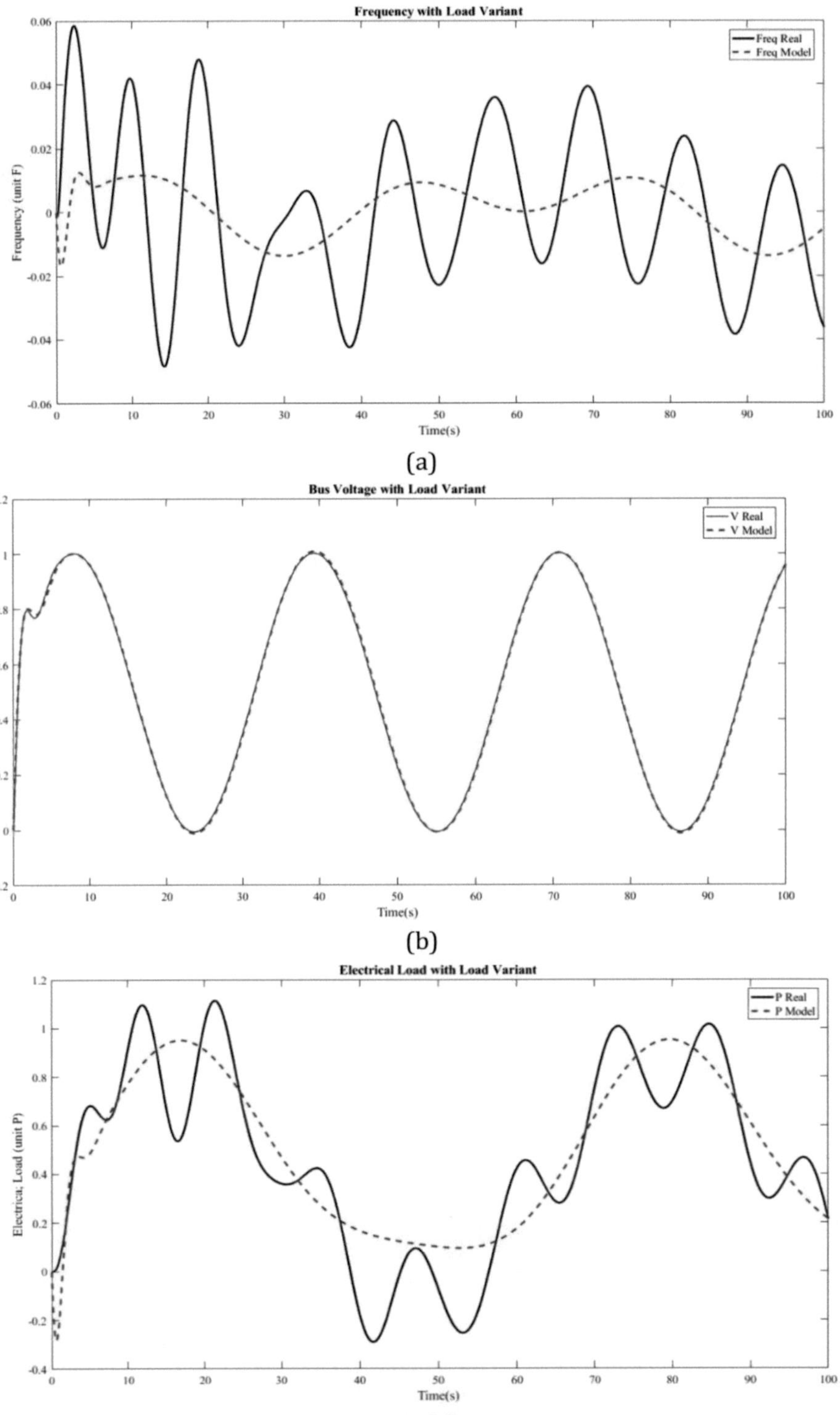

Figure 6.
Simulation results of cluster 1 load, (a) frequency, (b) bus voltage, (c) electrical power.

DOI: http://dx.doi.org/10.5772/intechopen.99126

2. Validate the following cluster 2 power generation system parameters (**Figure 7**):

3. Validate the following cluster 3 power generation system parameters (**Figure 8**):

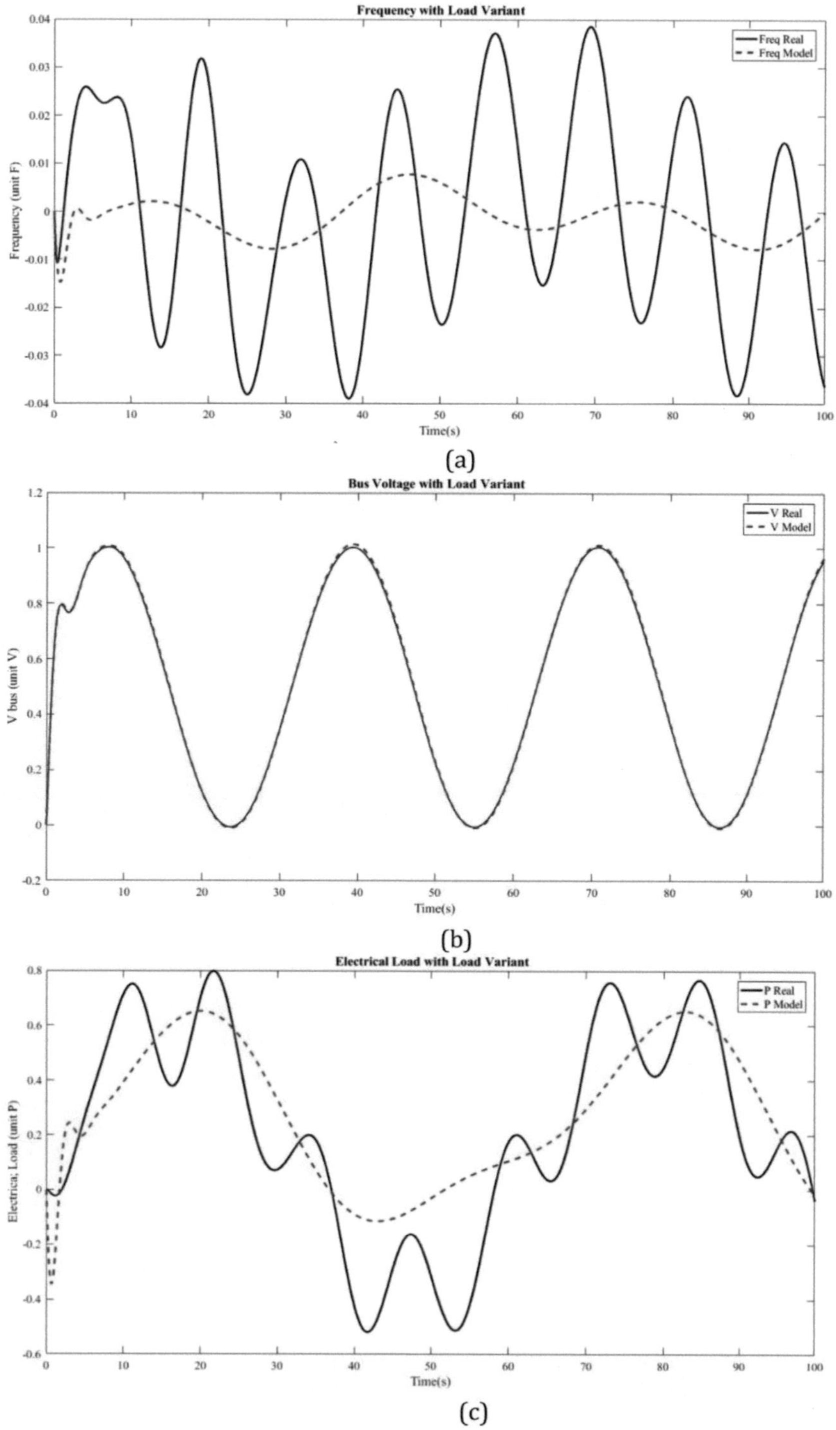

Figure 7.
Simulation results of cluster 2 load, (a) frequency, (b) bus voltage, (c) electrical power.

4. Validate the following cluster 4 power generation system parameters (**Figure 9**):

Comparative representation between the actual model and the validation model is presented in **Table 2** form as follows:

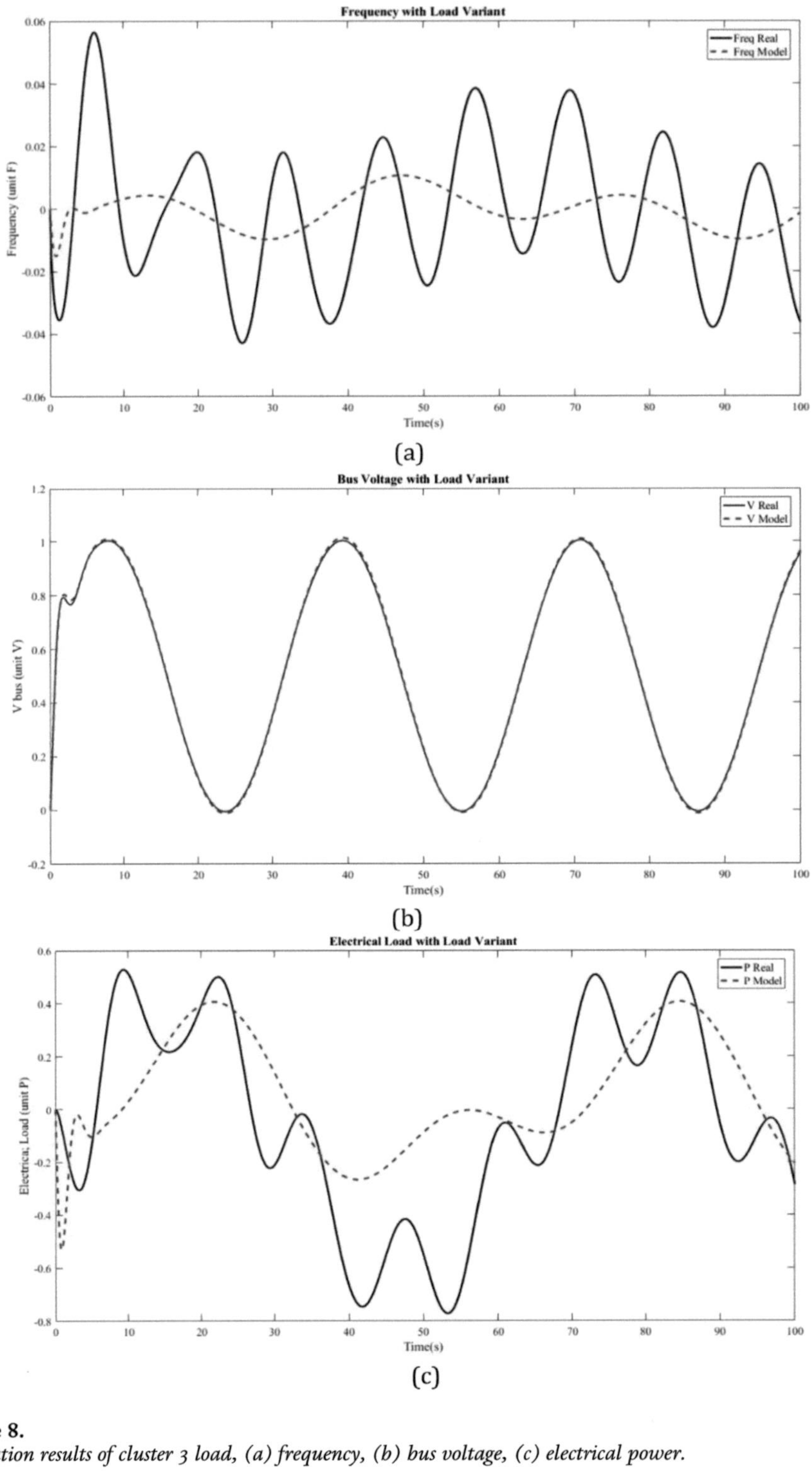

Figure 8.
Simulation results of cluster 3 load, (a) frequency, (b) bus voltage, (c) electrical power.

Based on the results obtained in **Table 2**, it can be seen that the IAE calculation is very significant between the actual output and the identification results, especially for changes in frequency and electrical power.

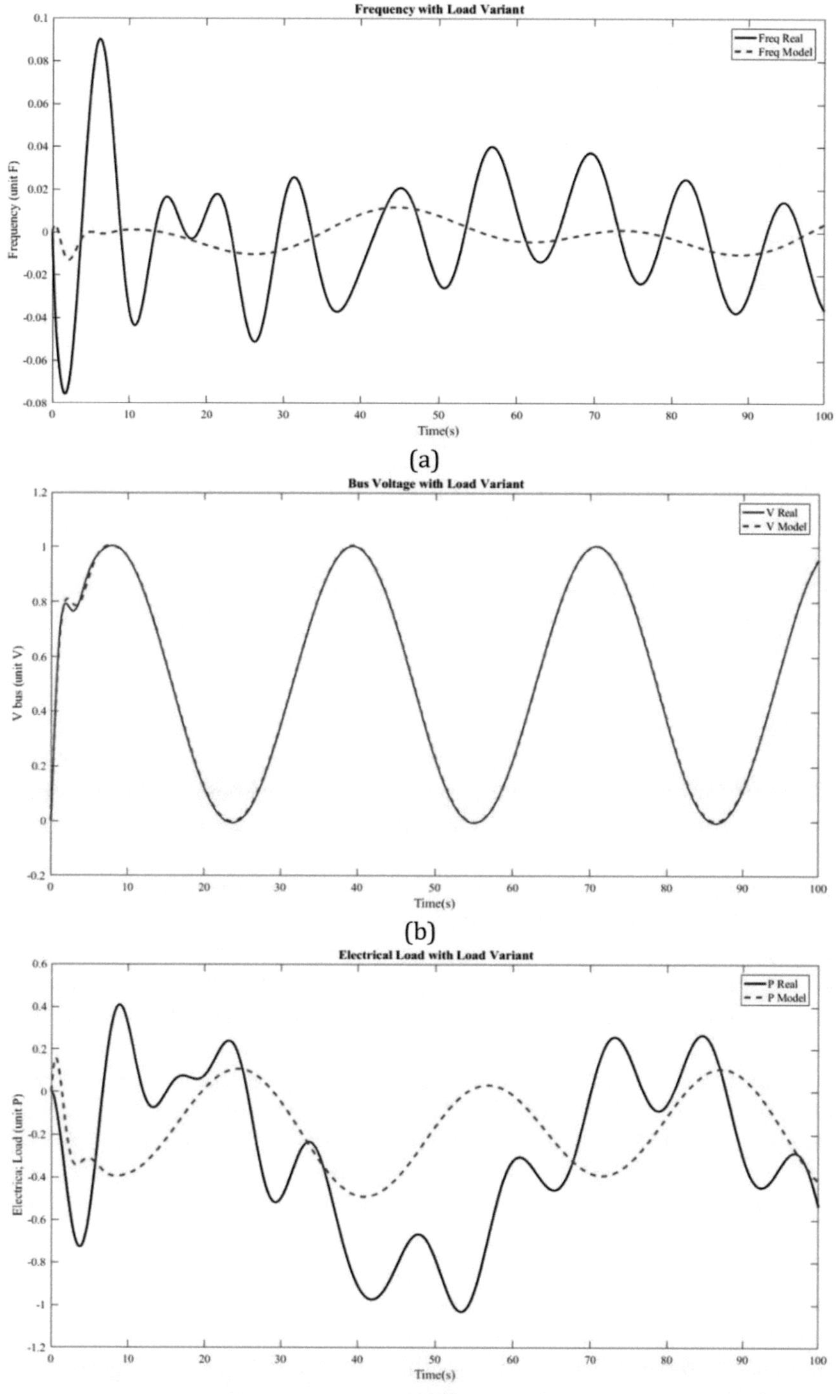

Figure 9.
Simulation results of cluster 4 load, (a) frequency, (b) bus voltage, (c) electrical power.

Kluster	IAE actual output			IAE identification model output		
	F	V	P	F	V	P
1	2.046	51.32	51.49	0.7489	51.27	50.54
2	1.856	51.32	35.81	0.3652	51.55	30.8
3	1.909	51.31	31.17	0.4764	51.46	17.48
4	2.184	51.3	37.7	0.4927	51.45	20.72
Average	1.999	51.313	39.043	0.5208	51.433	29.885

Table 2.
Comparison of integral absolute error (IAE) identification model output and actual output.

To calculate the performance comparison between the actual results and identification, the Mean Absolute Percentage Error (MAPE) formula is used as follows:

$$MAPE = \frac{\sum_{t=1}^{n}\left|\frac{Z_t - \hat{Z}_t}{Z_t}\right|}{n} \times 100\% \tag{11}$$

where Z_t and $\hat{Z}_t$ are the actual value and the identification value, while n is the number of calculation data sett.

Based on the MAPE calculation, the average frequency value is 73.95 percent, nominal voltage is 0.23 percent, and electric power is 23.46 percent.

4. Conclusion

The identification process of the power generation system in this paper is very supportive for further research, especially in the field of controlling the power generation system.

Prediction Error Minimization is very helpful in the identification process that is able to adjust the response to the desired signal model. Response time delay is calculated so that the generation system equipment on the mechanical and electrical side can work optimally. The delay time in **Figure 4** is included so that the power generation system equipment on the mechanical and electrical side can work optimally. The simulation results are obtained with better performance between the actual output and the identification model, namely the calculation of Integral Absolute Error (IAE), with MAPE for the average frequency value of 73.95 percent, nominal voltage of 0.23 percent, and electric power of 23,46 percent.

For future research work related to the identification of parameters of multi-engine generating systems as well as interconnection systems, it is necessary to carry out experiments. Even though it is in the form of a simulation, this work will help researchers to get closer to solving problems in real conditions.

Finally, this paper can be used as a reference for further research on the identification of power generation system parameters. In maintaining the stability of the frequency value on the mechanical side, the nominal voltage of the generator terminal and electric power when the electric load fluctuates.

Author details

Ismit Mado
University Borneo Tarakan, Tarakan City, Indonesia

*Address all correspondence to: ismitmado@borneo.ac.id

References

[1] Fruk, M., Vujisić, G., & Špoljarić, T. (2013, May). Parameter identification of transfer functions using MATLAB. In *2013 36th International Convention on Information and Communication Technology, Electronics and Microelectronics (MIPRO)* (pp. 571-576). IEEE.

[2] Chaves, W. D. S., Kaneko, E. H., Mollon, M. F., Niro, L., Vargas, A. D. N., & Montezuma, M. A. (2017). Parameters identification of a direct current motor using the trust region algorithm. *International Journal of Advanced Engineering Research and Science*, *4*(12), 237343.

[3] Amiri, M. S., Ibrahim, M. F., & Ramli, R. (2020). Optimal parameter estimation for a DC motor using genetic algorithm. *International Journal of Power Electronics and Drive Systems*, *11*(2), 1047.

[4] Fajardo, M., Guzmán, J., Vargas, W., Cepeda, J., & Urquizo, J. The Identification of Speed Regulation Systems Parameters for Hydraulic Turbogenerators.

[5] Soeprijanto, A. (2012). Desain Kontroller untuk Kestabilan Dinamik Sistem Tenaga Listrik.

[6] Robandi, I. (2006). Desain sistem tenaga modern. *Penerbit ANDI, Yogyakarta*.

[7] Ljung, L. (1995). *System identification toolbox: User's guide*. Natick, MA: MathWorks Incorporated.

[8] Law, A. M., & Kelton, W. D., (2000). *Simulation modeling and analysis* (Vol. 3). New York: McGraw-Hill.

[9] Box, G.E.P., Jenkis, G.M., & Reissel, G.C. (1994). Time Series Analysis Forecasting and Control. 3th Edition, Englewood Cliffs: Prentice Hall.

[10] Wei, W. W. S. (1990). Time Series Univariate and Multivariate Methods. Canada: Addison Wesley Publishing Company, Inc.

[11] Walpole, R. E. (1990). Pengantar statistika, edisi ke-3 (Introduction to statistics). Penerbit PT Gramedia Pustaka Utama.

[12] Wiyono, B. B. (2001). Statistik Pendidikan: Bahan Ajar Mata Kuliah Statistik. *Malang: FIP UM*.

[13] Anderson, P. M., & Fouad, A. A. (2008). *Power system control and stability*. John Wiley & Sons.

[14] Demello, F. P., & Concordia, C. (1969). Concepts of synchronous machine stability as affected by excitation control. *IEEE Transactions on power apparatus and systems*, *88*(4), 316-329.

[15] Aswathi, C. R., & Jacob, J. (2016, May). Power system stabilizer for single machine infinite bus system for robustness against inertia constant variations. In *2016 International Conference on Advanced Communication Control and Computing Technologies (ICACCCT)* (pp. 467-472). IEEE.

[16] Fortulan, R. L. V., & Alberto, L. F. C. (2019, August). Transient Stability Analysis of a Single Machine Infinite Bus System with Uncertainties in Generated Power. In *2019 IEEE Power & Energy Society General Meeting (PESGM)* (pp. 1-5). IEEE.

[17] Aghababa, M. P., Marinescu, B., & Xavier, F. (2020). Observer-based Tracking Control for Single Machine Infinite Bus System Via Flatness Theory. *International Journal of Electrical and Computer Engineering (IJECE)*.

[18] Mado, I., & Soeprijanto, A. (2014, August). Design of robust-fuzzy

controller for SMIB based on power-load cluster model with time series analysis. In *2014 Electrical Power, Electronics, Communicatons, Control and Informatics Seminar (EECCIS)* (pp. 8-15). IEEE.

Chapter 3

Particle Filter Based Approach for Remaining Useful Life Prediction of High-Speed Shaft Bearing in Wind Turbine Generators

Sharaf Eddine Kramti, Jaouher Ben Ali, Hugo Andre, Eric Brhhoefer and Mounir Sayadi

Abstract

This work involves a novel data-driven procedure using vibration analysis for bearing health prognosis. In this work, we investigate the time-domain features and applying spectral kurtosis features in order to extract the damage indicators which eventually represent the degradation of the high-speed shaft bearing (HSSB). These damages were characterized by their Monotonicity, Trendability, and Prognosability. The most appropriate indicator was then used as a health index for the remaining useful life (RUL) prediction task. In this study, we used an integrated approach based on Particle Filter approach which was then developed for direct RUL prediction of HSSB. This methodology was validated using real world vibration data wind turbine gearbox. The experimental results and the prognostics metrics like fitness degree equal to 0.9941 shown that the Particle Filter approach is more feasible prediction tool.

Keywords: Bearing vibration monitoring, particle filter, prognostics and health management, Wind turbine generator (WTG)

1. Introduction

The economic development was based on low cost energy. The growth of renewable energy, especially that is produced by wind turbine generator (WTG) helps continued growth while curbing CO_2 emission. While the costs of WTG power production is low, operations and maintenance costs are higher than expected due to higher rates of electrical and mechanical failures. According to national renewable energy laboratory (NAREL), failure in HSSB accounts for 48% of all drive-train damage. Mechanical failure in (WTGs) leads to sudden downtime and electricity production cessation causes a higher than expected maintenance cost. Prognostics and health management (PHM) of WTGs aim to predict the future behavior of the generator's health condition by estimating the RUL of HSSB. This allows requires a proactive maintenance policy that can reduce the operations and maintenance cost and provides better balance of plant.

Mechanical failures in WTG bearings drive train is often the result of moisture contamination of the oil, especially in offshore application. Contamination causes reduction in oil lubrication, reducing the life cycle [1]. Bearing failures can create long down time and costly maintenance. To maintain balance of plant, implementation of a Condition Based Maintenance (CBM) program should be considered. CBM along with PHM are able to identify the failure and estimate the RUL of the elements.

PHM is a maintenance paradigm that merges diagnostics, future loads and a damage propagation model to estimate the RUL. Diagnosis describes the current state and the damaged component in the system. Applying the current state to a propagation model, based on future estimated load, and a point where maintenance is appropriate (e.g. the threshold), the RUL [2–5] can be estimated. As shown in **Figure 1** the PHM cycle is composed of three principal parts ranged as follows:

- Observe contains two steps: Acquire data from sensors installed on the machine, then data processing/feature extraction to generate a condition indicator (CIs). Some other techniques, such as digital signal processing improve the signal to noise and quantify the representative statistics of component damage.

- Analysis is a three steps process. First, a condition assessment is made of the observed CIs. Second, the given diagnostic represents is given representing the component state of health. This also includes fault isolation and identification. Third, using the current state of health and the threshold, estimate the RUL and the associated confidence level of the RUL for the given component.

- Act is a two-step process. First, decision support tool evaluates the evidence, which can generate a maintenance intervention. The last step is Human Machine interfaces, which display the PHM status.

A PHM architecture [6, 7] leads to implement a paradigm that supports maintenance planning with the goal of eliminating unscheduled maintenance and improving operational readiness.

The activity associated with PHM is growing, with a number of scientific papers and several reviews have been published in CBM domain. Lee et al. [4] introduce different strategies for the detection of failures associated with rotating equipment.

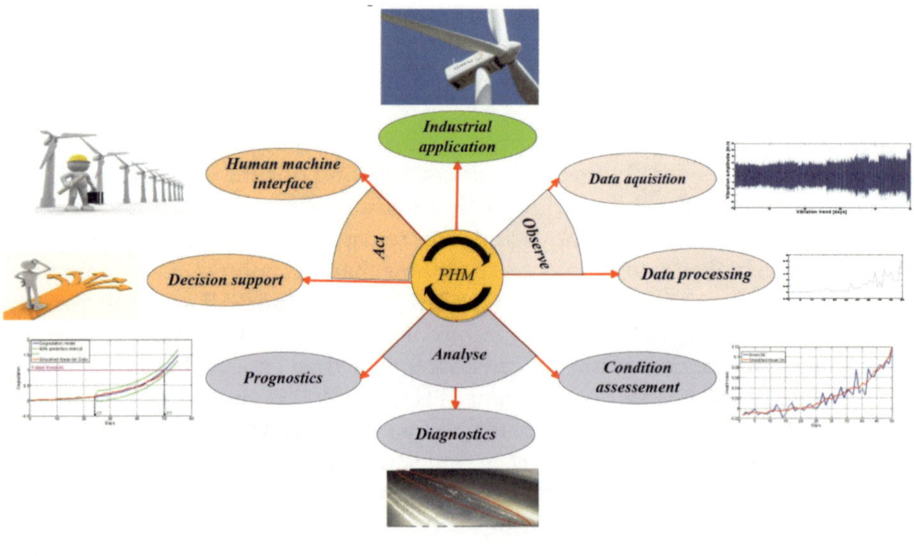

Figure 1.
PHM cycle retyped and adapted from the ISO 13374.

Lee goes on to describe many of the analysis algorithm that can be used, and the analysis performance (advantages and disadvantages).

The review by Jardine et al. in [8] is an excellent overall summary of CBM. Even though this publication is old, all the proposed approaches used in data processing and maintenance decision support are still valid until now.

In [9], the authors applied a Kalman smoother approach with confidence bounds in order to predict the RUL of HSSB. The prognostic approach is hybrid using both a data driven approach and physics-based degradation model. Vibrating data collected over 50 days was used to estimate the damage associated with propagating inner race crack. Using a Kalman filter, the unknown parameters associated with Paris' Law model were determined in order to estimate RUL.

In [10], Kramti et al. applied Elman neural network (ENN) technique. They proposed an approach using time-domain features and frequency-domain via Spectral Kurtosis (SK) as inputs. These features were extracted from raw vibration bearing signal in order to predict the RUL of HSSB. The architecture of an ENN is built with two hidden layers. The first hidden layer is composed of five neurons, and the second hidden layer is composed of three neurons. The hidden layers transfer function is Logarithmic sigmoid. The output layer is composed of one neuron using a pure linear transfer function ranging between 0 and 1. This ENN gave a prediction horizon of 20 days. While a powerful model, there were some gaps in the predicted RUL when compared to the true RUL with large fluctuations.

The authors in [11] used a support vector regression (SVR) approach based on classical, time domain features. This SVR model used spectral kurtosis (SK) to predict the RUL of HSSB. Spectral kurtosis derived indices reduce the noise by using the Short-Time Fourier Transform (STFT) and provide good trendability and monotonicity metrics.

In the SVR study, the model was trained by 60% and tested by 40% of area under SK index data. In other words, it was trained using 40% of the data and tested using 60% of the same type of data used in the first step. The experimental results have shown that the estimated RUL based on area under SK index tracks the actual RUL with a small prediction error using 60% of training data. Unfortunately, the model did not predict any SK values.

The [12], the authors used Acoustic Emission (AE) data which were delivered from four bearings. The fault feature was Root Means Square (RMS) and Signal Intensity Estimator (SIE). All bearings features were fitted. The bearing feature bearing 1, were used as inputs of training process of three learning machines using: Gaussian Process Regression (GPR), support vector machine regression (SVMR) and multilayer artificial neural network (ANN). The other three bearings features are used as input of process test. The experimental results have shown that ANN gave the lowest error compared to SVMR and GPR.

The proposed method aims to improve the reliability of the HSSB and reduce maintenance cost. Therefore, the new proposed failure prognostics method is based on the analysis of the behavior of vibration signal. The proposed failure prognostics method uses the most suitable feature, which is then mapped to a Health index (HI). The RUL is then the estimated time for the current HI, to the HI threshold were it is appropriate to do maintenance.

The remainder of this work is organized as follows. Section 2 reports the bearing characteristic and describes the data used in the experiment. In Section 3 the proposed methods and techniques used like features extraction methodology, particle filter prognosis is introduced with the model degradation. In Section 4 we detail the procedure to obtain the RUL. Section 5 A discussion and a comparison with some previous work and methodologies in the literature. Finally, we conclude this chapter and future work are synthetized in Section 6.

2. Experimental steps

The vibration signals data were obtained from an online condition monitoring system from Green Power Monitoring System (GPMS). During 50 days of measurements, data was collected at a rate of once per 10 minutes (144 acquisitions per day), where once per day raw vibrating signal data was downloaded. The sensor monitoring the high-speed bearing was sampled at 97656/second for six seconds, along with tachometer data. The data was collected from a real wind turbine generator (S88, Suzlon) with 2 MW electric power generation.

The failing bearing supports the high-speed shaft, which drives the generator. This shaft rate was approximately 1800 revolutions per minute (these are doubled feed induction machines, so that the gearbox/generator is not synchronous at 60 Hz, while the output of the generator is electronically controlled to be line synchronous).

The vibrating signal data given by the Green Power Monitoring System from USA. The vibrating sensor was made by MEMS accelerometer (analog device ADXL001, the bandwidth is a 32 kHz, the resonance at 22 kHz, with a sensitivity of +/−70 Gs. The data was sampled by a delta-sigma analog to digital converter ADS1271 (24 bit). The vibrating sensor was installed above the HSSB.

On the fiftieth day, the HSSB was inspected, where the damage to the inner race fault was verified, as indicated in **Figure 2**. The HSSB model is 32222-J2 [13] tapered roller bearing it is made by SKF. Bearing dimensions are 200 mm in outside diameter; the inside diameter is 110 mm, the width is 56 mm, it has a 20 rolling elements each one has 46 mm of width, the taper angle is 16°, weighs 7.10 kg and the speed-limit is 3200 r/min.

The rotating speed of WTG depends on climatic condition, the pressure altitude, offshore or onshore location. Indeed, these environmental conditions directly influence the bearings radial and axial loads. When wind speed is near the cut-in wind (i.e. the lowest wind speed needed for power generation), the main shaft will be in lower speed (high speed shaft perhaps 1500 rpm). When the wind speed is higher than cut-in wind, the main shaft speed operates higher on the power curve, close to 1800 RPM, the **Figure 3** shows the mean of speed shaft over 50 days. Loads are moderated by the wind turbine blade pitch. **Figure 4** shows the concatenation of the 50 days of measurements data in a 6-second period each time. It is clear that the RMS increases with time and damage propagation.

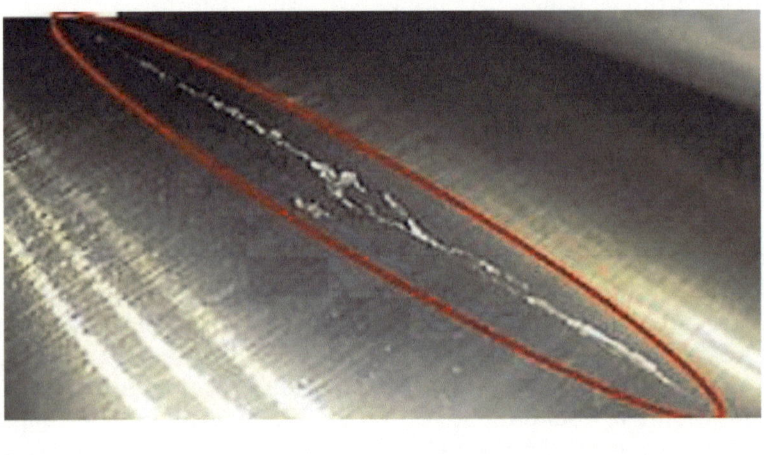

Figure 2.
Inner race fault.

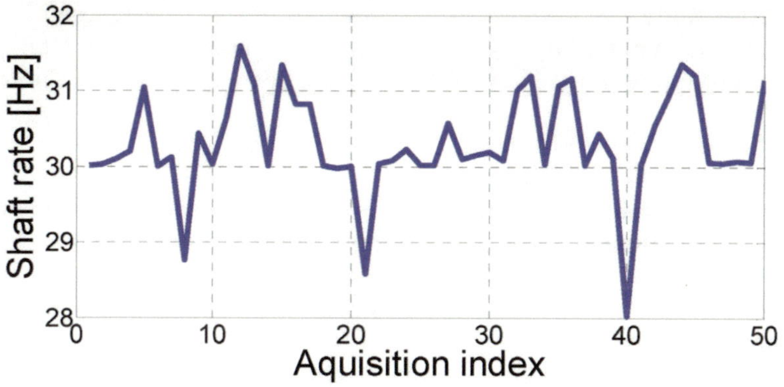

Figure 3.
The mean speed variation in high speed shaft over 50 days.

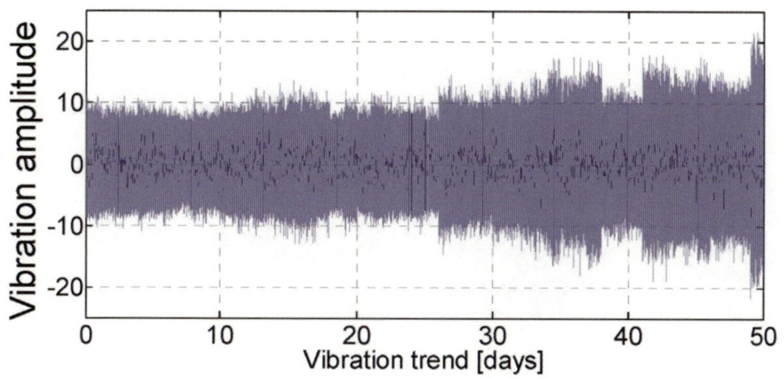

Figure 4.
The historical vibration data ending with inner race fault.

A. Proposed prognostic approach

In this chapter, we propose an effective prognostic method to estimate degradation in the bearing. The proposed method for HSSB failure is summarized in **Figure 5**. The proposed method consists mainly of three task features definition, feature selection and RUL prediction. During the first task of the proposed method, two types of features are used to extract information from vibration data: classical time-domain features and a SK feature (which is a frequency-domain feature). Once the features definition task has been done, feature selection was performed. In this task, we use three metrics in order to determine a suitable feature, which will inform the RUL prediction step. Finally, the RUL prediction step implies the use of the particle filter method to predict the RUL according to prognostics metrics.

B. Features

1. The Classical features

Classical features are presented in **Table 1** [5, 14] were applied on historical vibration signal run-to-failure over 50 days as shown in **Figure 6(a)**.

Generally, time-domain features are well proven and historically, are the basis of many condition-monitoring systems. The classic features used in this study are; RMS, Kurtosis, Skewness, Peak to peak, Crest Factor, Mean, Standard deviation (Std), Energy, and Entropy. A detailed description of the classical features are given as follows in **Table 1.**

2. Features derived from SK

In order to estimate the RUL process, it is necessary to identify features that show trendability, monotonicity and that can be used as a surrogate for component damage. For this study, we used features derived from SK. SK typically involves the band-pass filtering of the raw data to remove signal that is not associated with the fault. Also, it improves the SNR.

The approach requires a band-pass filter to find a bandwidth that emphasizes the demodulated impulsive signature (associated with bearing fault) which is hidden in the raw vibrating signal. For a full treatment on SK, pleases refer to Randall and Antoni [15].

The kurtosis is defined as:

$$K = \frac{\frac{1}{N}\sum_{i=1}^{N}(x_i - \overline{x})^4}{\left(\frac{1}{N}\sum_{i=1}^{N}(x_i - \overline{x})^2\right)^2} \tag{1}$$

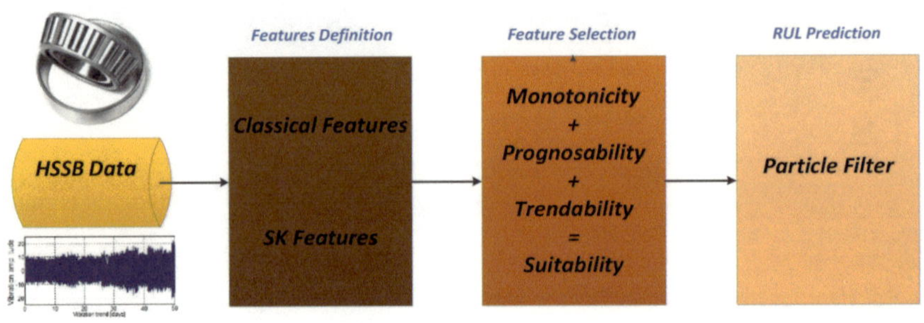

Figure 5.
Flow chart of the proposed prognosis method.

Feature name	Mathematical expression
RMS	$\left(\frac{1}{N}\sum_{i=1}^{N}x_i^2\right)^{\frac{1}{2}}$
Kurtosis	$\frac{1}{N}\sum_{i=1}^{N}\frac{(x_i-\overline{x})^4}{\rho^4}$
Skewness	$\frac{1}{N}\sum_{i=1}^{N}\frac{(x_i-\overline{x})^3}{\rho^3}$
Peak to peak	$x_{\max} - x_{\min}$
Crest Factor	$\frac{x_{\max}}{RMS}$
Mean	$\overline{x} = \frac{1}{N}\sum_{i=1}^{N}x_i$
Standard deviation	$\sigma = \left(\frac{1}{N}\sum_{i=1}^{N}(x_i - mean)\right)^{\frac{1}{2}}$
Energy	$\sum_{i=1}^{N}x_i^2$
Entropy	$-\sum_{i=1}^{N}x_i \log(x_i)$

Table 1.
The proposed features.

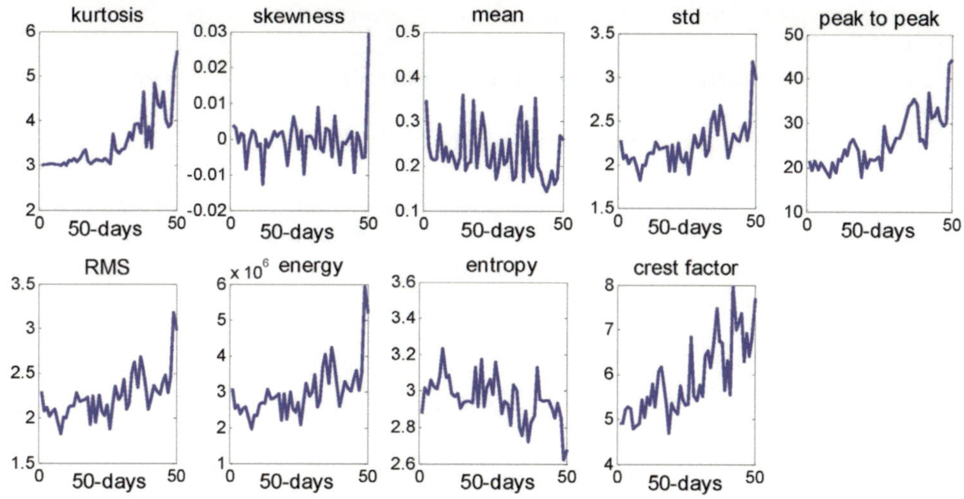

(a) Time Domain Features

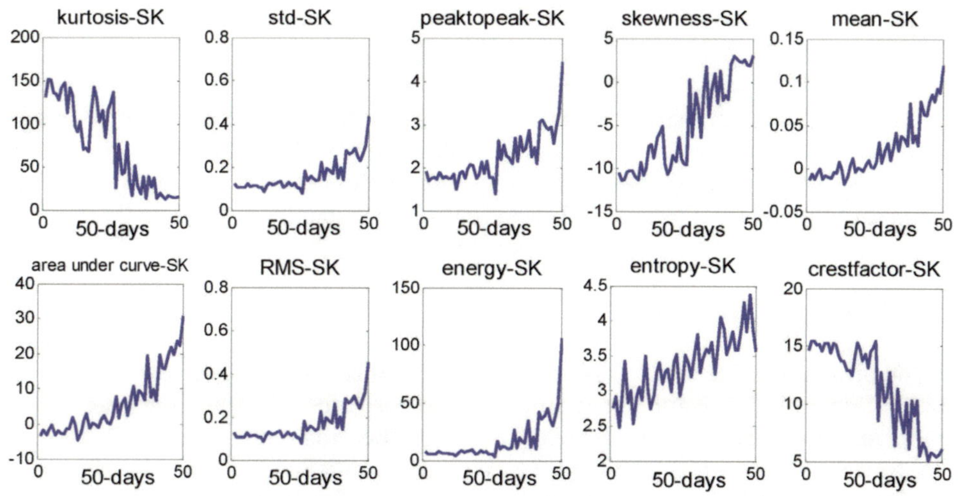

(b) SK Features

Figure 6.
The trend for the 50 days (a) representing the time domain, classical features, and (b) features derived from SK.

Where i is the sample index, x and $\bar{x}$ are the sample time index and sample mean respectively and N is the number of samples. This normalized fourth moment is defines the "peakedness" of the signal. The spectral kurtosis of signal is described as the kurtosis of its spectral elements. The SK is defined as follow [16].

$$SK(f) = \frac{\langle |X^4(t,f)| \rangle}{\langle X^2(t,f) \rangle^2} - 2 \tag{2}$$

Where $X^2(t,f)$ and $X^4(t,f)$ are the second-order and fourth-order cumulate respectively of a band-pass filtered signal of $x(t)$ around f. ‹•› correspond to the time frequency averaging operator.

The most important characteristics of this description defined as:

- In case of stationary system, the SK is a permanent function of frequency.
- In case of stationary Gaussian system, the SK is the Gaussian process.

The SK has been shown to be a more powerful indicator of damage than raw signal kurtosis. SK detects the high-frequency train of impulse derived from a damaged bearing. It can be shown that the SK of a non-stationary system $x(n)$ and a damaged stationary noise $b(t)$ source is defined as

$$SK_{(x+b)}(f) = \frac{SK_x(f)}{(1+\rho(f))^2} + \frac{\rho(f)^2 SK_b}{(1+\rho(f))^2} \tag{3}$$

Where $f \neq 0$, then $\rho(f)$ is the signal-to-noise ratio (SNR) as function of frequency. If b (t) is an additive stationary Gaussian noise independent of $x(t)$, then the SK is transformed into.

$$SK_{(x+b)}(f) = \frac{SK_x(f)}{(1+\rho(f))^2} \tag{4}$$

Now, it can be seen that the SK is able to characterize and detect bearing damage that is masked by a non-stationary signal in the frequency domain. Additionally, this shows that the when using the SK, for a nominal, stationary signal, the value of SK is approximately 0. For non-Gaussian, damaged signal (e.g. transients), the value is $\neq 0$, see **Figure 7** for more explication.

3. HSSB degradation model

The principal cause of damage and early bearing failures are overload, inadequate lubrication, lubrication contamination, or corrosion. Bearing faults can appear on inner race, outer race, rollers and cage. Bearing damage may result in large, quasi-periodic impacts, which degrade exponentially over time [17, 18].

When the model degradation with a defined level of damage, the measured data can be used to calibrate and identify the parameters of model. If the model parameters are known, they can be applied to estimate the prospective behavior of the damage.

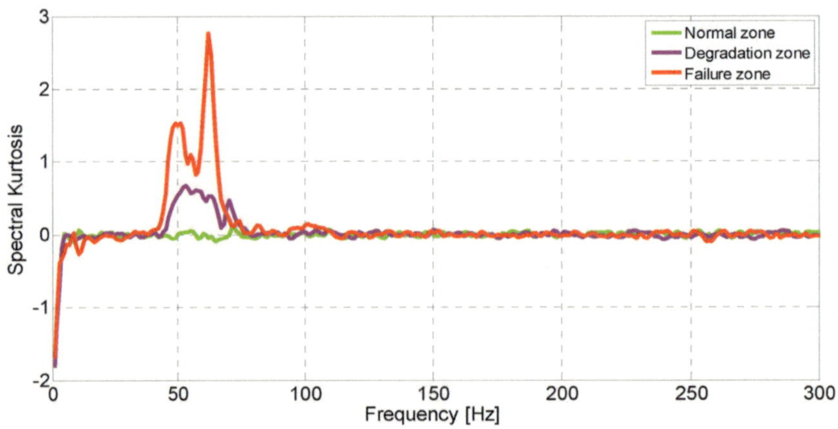

Figure 7.
Sate of health in HSSB using spectral kurtosis.

- The model degradation is defined as

$$d = a \exp{(bt)}^2 \tag{5}$$

Where d is the magnitude degradation, a, b are the model constant parameters and t is the time index. This model allows a “best fit” trend of the health index of HSSB.

C. Particle filter based prognosis

In this section we present a brief review of Particle Filter (PF). More detailed study on PF can be found in [19] In this chapter we present only the basic theory.

PF has been applied in numerous engineering domain such as robotics, aerospace, automatic control, etc. and more currently in diagnosis and prognosis. PF may also be known as the sequential Monte Carlo method. The main uses of PF are to accurately model the degradation state with a set of particles. The PF has a corresponding state values, and a correlated set of particles weights, which correspond to the discrete Probability masses of the distinctive particles.

In PF, the Bayesian update is processed in sequential mode with samples (or particles) having the information probability of hidden parameters: when a new measured data is obtained, the posterior step is used as the information for the present step, and the parameters are updated by multiplying it with the likelihood function.

The particle can be created and updated recursively by the use of non-linear state-transition model, illustrating the evolution of the system under control. The Bayesian tracking task is described by two equations as follows

- The state equation, the state transition function f.

$$x_k = f(x_{k-1}, \theta_k, v_k) \tag{6}$$

- The model observation, measurement function h.

$$z_k = h(x_k, \omega_k) \tag{7}$$

Where k is the time step index, x_k is the system state at the preceded step in this work x_k represents the damage state of HSSB. θ_k is a model parameters vector, z_k is measured data, ω_k and v_k are respectively measurement data and process noise. All these variables fluctuate at each time step, and the progress from the *k-1* to the *k* step is produced by the transition function f. As mentioned before, this work is about prognostics area so the state transition function f is designed as damage model.

According to the state model in (Eq. (6)), the HSSB degradation model in (Eq. (5)) can be edit in the following shape

$$x_k = x_{k-1} \exp{(b_k \Delta t)}^2 \tag{8}$$

Where the process noise v_k is neglected because it can be managed through the uncertainty in model parameters. In case of measurement function, it is supposed that z_k is the selected feature that reflects the HI of HSSB (see **Figure 8**) This feature includes measurement Gaussian noise $\omega_k \sim N\,(0,\sigma)$ which is applied with unknown standard deviation σ. Consequently the unknown parameters are $\theta = [b,\sigma]$, containing the damage state x_k which is acquired based on the model parameter b_k.

A probability density function (PDF) $p(x_k|z_{1:k})$ is needed to obtain the distribution of the possible states of x at time k. The initial state is estimated by the state distribution $p(x_0|z_0) = p(x_0)$. It is assumed that the PDF is known. The optimal Bayesian solution is given by iterating prediction and update functions respectively:

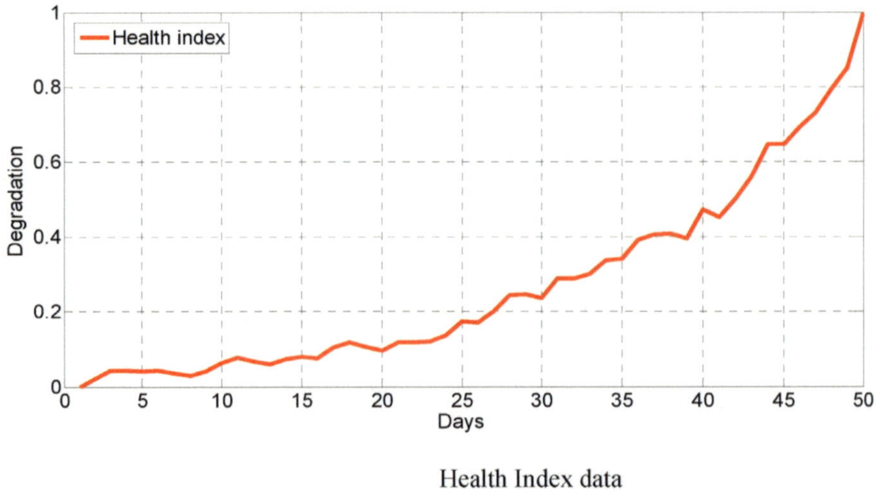

Figure 8.
Health index data.

- Prediction

$$P\langle x_k|z_{1:k-1}\rangle = \int P\langle x_k|x_{k-1}\rangle P\langle x_{k-1}|z_{1:k-1}\rangle dx_{k-1} \tag{9}$$

- Update

$$P\langle x_k|z_{1:k}\rangle = \frac{P\langle z_k|x_k\rangle P\langle x_k|z_{1:k-1}\rangle}{P\langle z_k|z_{1:k-1}\rangle} \tag{10}$$

Unfortunately, this solution cannot be obtained systematically, but PF is a robust approach designed to obtain an approximate solution via feedback.

Figure 9 shows an algorithm of the PF which can be recapitulated as follow [20, 21].

The primary step consists in subdividing the initial state distribution $p(x_0)$ into n samples called also particles. The next three steps are then reiterated until the appropriate results are achieved see **Figure 9**.

1. ***Prediction***: in this step the particles are generated through the state model from *k-1* to the *k* step providing each incremental time a new PDF. The information from preceding step should be fully accessible.

2. ***Update:*** the model parameters and state degradation are updated. The measurement data are utilized to computing the likelihood function $p(z_0|x_k)$, and provides weights to the particles. The new particles states converge to the actual one provided by the last measured data which have the higher likelihood, as they are more suitable to describe the system state.

3. ***Resampling:*** the concept of resampling is to keep all samples (i.e., particles) have the same weight. Particle that have a low weight are removed. And particles with the high weights are manifold. This operation can favorite the best filter results. Namely a higher number of particles with poor weight can degenerate the filter results in the previous stages.

All these steps are used during the learning process. In case there is no available data, no measured data z_k should be used to compute the likelihood function, and

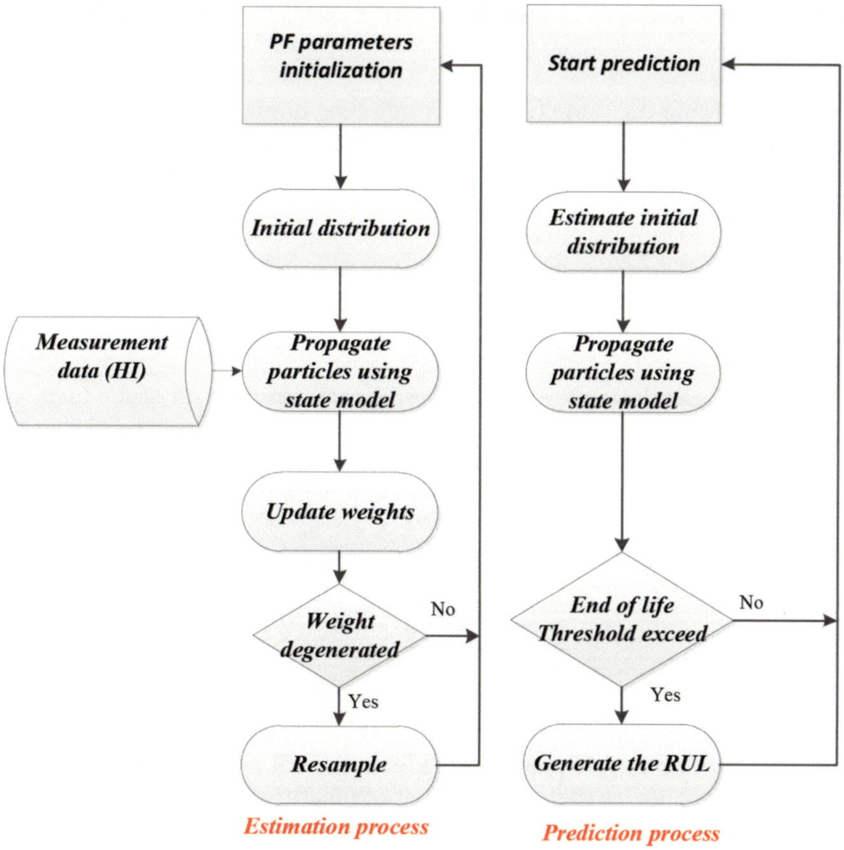

Figure 9.
Illustration of PF principle estimation process and prediction process.

the prognosis pass into prediction phase. In this prediction step, the particles are diffused via the state model. When the failure threshold is crossed the latest distribution of particles are the most appropriate state, (i.e., the state expressed via the weighted average of the particles).

3. Tracking fault degradation

A. Features definition

In order to define the degradation of HSSB, two types of features have been used [5]. The first is time domain indices of classical features such as: kurtosis, skewness, mean, standard deviation (std), peak to peak, root means square (RMS), energy, entropy and crest factor. The second are features based on the SK but operated on by the same classical features: kurtosis-SK, skewness-SK, mean-SK, standard deviation-SK (std-SK), peak to peak-SK, root means square-SK (RMS-SK), energy-SK, entropy-SK, crest factor-SK and area under curve-SK which is added. These features are extracted for the 50 day data set, as shown in **Figure 6**.

B. Feature selection

Feature selection performed to identify the best analysis for classification and prognosis. The principal idea is find the most relevant features that provide useful information and that can predict a future state. This was done by two methods:

- By transforming the same feature set into new domain, such as main component evaluation or linear discriminant evaluation [4]. This approach unfortunately generates new feature sets in the new domain that are dissimilar to actual features.

- By choosing a feature that allows for prognosis. As noted in recent literature on prognosis, metrics such as: trendability, prognosability and monotonicity [22, 23] can be used to determine the most appropriate feature set.

Monotonicity can define the main negative or positive trend of the feature. This is a powerful metric to detect degradation because degradation in bearing is an irreversible and a growing process. The monotonicity of a group features is affected by the mean between the number of positive and negative step for every assessment point of time. Suppose that n is the number of assessment point of time, the monotonicity will be defined as follows

$$Mono = \left| \frac{noof\left(\frac{d}{dt} < 0\right) - noof\left(\frac{d}{dt} > 0\right)}{n-1} \right| \tag{11}$$

The range of monotonicity value between 0 and 1, non-monotonic features will take the value of 0 and greatly monotonic features will take the value of 1.

Trendability quantifies the correlation of the features vs. time. If the feature is constant, the correlation with time will be 0. However, if the derivative of the feature is linear, the correlation with time will take on a non 0 value. In the same way, correlation can change with increase in non-linearity (i.e, a nonlinear feature will result in low correlation). Trendability is ranged between -1 and 1 it is defined as follows

$$\begin{aligned} Tren &= |corrcoef(time, feature)| \\ &= \frac{n\left(\sum timefeature\right) - \left(\sum time\right)\left(\sum feature\right)}{\sqrt{\left[n\sum time^2 - \left(\sum time\right)^2\right] - \left[n\sum feature^2 - \left(\sum feature\right)^2\right]}} \end{aligned} \tag{12}$$

Prognosability is the exponential of the standard deviation of degradation feature measure divided by the difference between final and first value of degradation feature measure.

$$Prog = \exp\left(-\frac{std(\text{deg}measures)}{mean(|final\text{deg}measure - first\text{deg}measure|)}\right) \tag{13}$$

According the numerical values of suitability (Eq. (14)) in **Tables 2** and **3**, the most significant feature for the prognostic task among all ones, is the mean-SK which has an exponential growth. Note the feature is usually combined with noise. The noise can obscure the trend and reduce the power of the RUL estimation.

$$Suitability = Monotonicity + Prognosability + Trendability \tag{14}$$

Extracted features are usually associated with noise. The noise with opposite trend can sometimes be harmful to the RUL prediction. In addition, one of the

Features	Monotonicity	Trendability	Prognosability	Suitability
Kurtosis	0.1020	0.8081	0.7810	1.6911
Std	0.1020	0.6555	0.6944	1.4519
Peak to peak	0.1020	0.8004	0.7599	1.6623
RMS	0.1020	0.6479	0.6892	1.4391
Skewness	0.0612	0.1007	0.7913	0.9532
Energy	0.1020	0.6319	0.7084	1.4423
Crest factor	0.0204	0.7852	0.7419	1.5475
Mean	0.0204	0.2496	0.5232	0.7932
Entropy	0.0204	0.5275	0.5393	1.0872

Table 2.
Classical features results during 50 days.

SK-features	Monotonicity	Trendability	Prognosability	Suitability
Kurtosis-SK	0.1020	0.8873	0.6469	1.6362
Std-SK	0.0612	0.7702	0.7957	1.6271
Peak to peak-SK	0.0204	0.7997	0.7988	1.6189
RMS-SK	0.0204	0.7928	0.7938	1.6070
Skewness-SK	0.0204	0.9024	0.6881	1.6109
Energy-SK	0.0612	0.7061	0.8316	1.5989
Crest factor-SK	0.0612	0.8896	0.6525	1.6033
Mean-SK	**0.1429**	**0.8852**	**0.7730**	**1.8011**
Entropy-SK	0.0612	0.8332	0.5920	1.4864
Area-under curve	0.0612	0.8851	0.7730	1.7193

Table 3.
Derived features-SK results during 50 days.

feature performance metrics, introduced above is not robust to noise. Therefore, a causal moving mean filter with a lag window of 5 steps is applied to the most suitable extracted features, where "causal" means no future value is used in the moving mean filtering. The **Figure 10** showing the mean-SK feature before and after smoothing, the smoothed mean-SK is used as health indicator of HSSB.

The smoothed mean-SK feature as shown in **Figure 8** is normalized in range [0 1]. 0 corresponds to 0% degradation and 1 corresponds to 100% degradation. This feature is used as Health index (HI) data in the following prognosis steps and the smoothed feature is normalized using (Eq. (15)).

$$mean - SK' = \frac{mean - SK - mean - SK_{\min}}{mean - SK_{\max} - mean - SK_{\min}} \tag{15}$$

4. RUL prognostic

A. RUL prognosis based particle filter approach

Measured Data is applied to predict model parameters, which are used then to estimate the RUL. The damage model equation in (Eq.(8)) is defined as follow: the

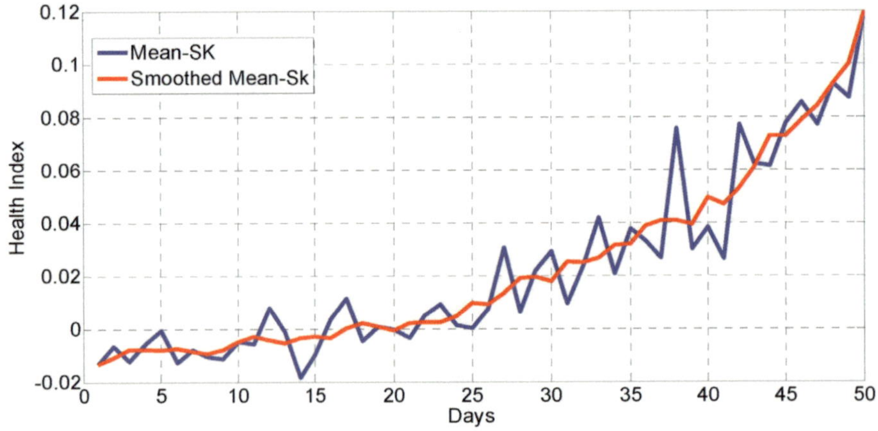

Figure 10.
The selected feature.

time interval Δt equal to 1 which corresponds to one inspection per day. Also, the model parameters b_k and the damage state at the previous step x_{k-1}, and the standard deviation of measurement error s. The initial distribution is $P\times Q$ matrix of probability parameters of the initial distribution, which P is the number of unknown parameters and Q is the probability parameters. In this case, the prior information are not available, it is supposed that the initial distribution of these unknown parameters **(P = 3)** are uniform, where the probability parameters are **(Q = 2)**, lower and upper bounds:

$$x_0 \sim U(0.01, 0.03), b_0 \sim U(0.038,\ 0.04), s_0 \sim U(0.251,\ 0.253).$$

The other setting for the prognosis using particle filter are the number of samples (or particles) n and significance level for adjusting the prediction interval **(PI)** and the confidence interval **(CI)**. In this work, it is used **(n = *5000 particles*)** and 90% of significance level. For more details according to the number of samples please read the work published in [19].

The other results can be plotted such as the model parameter and the prediction of failure state can be obtained by the use of sampling results during the updating process. The sampling results can be displayed for any variable at each step. In this work the exact values of the model parameters are known, the result should be compared with the known values. The exact value of **b = 0.03919** and **s = 0.2520**, the failure state can be calculated using the (Eq. (8)). More graphical results are shown in the following plots: **Figures 11–13**.

Once the model parameters are classified as a physisc based approach, the mathematical exponontial fucntion in (Eq. (8)) is trained using a the derived data. The particle filter uses these as imputs to predict the remaining time until the degradation propagates to a maintenance threshold.

The expremental results were made using a total of 50 measurement data points for HSSB HI (see **Figure 8**). The Particle filter process is runing only with 48% of measured data (24 points) and the rest of degradation trend were predicted by the proposed method.

The main goal of the particle filter based prognosis is to predict the degradation behavior by the use of the exponential degradation model. If the model accuratly represents the HI, then it can be appliyed to find the RUL using (Eq. (16)). The true RUL is achieved by substracting the current time (24 days) from the failure

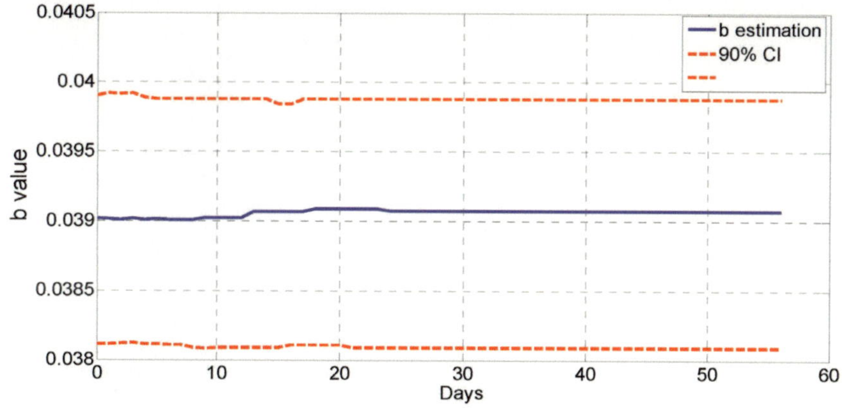

Figure 11.
Estimation of b model parameter.

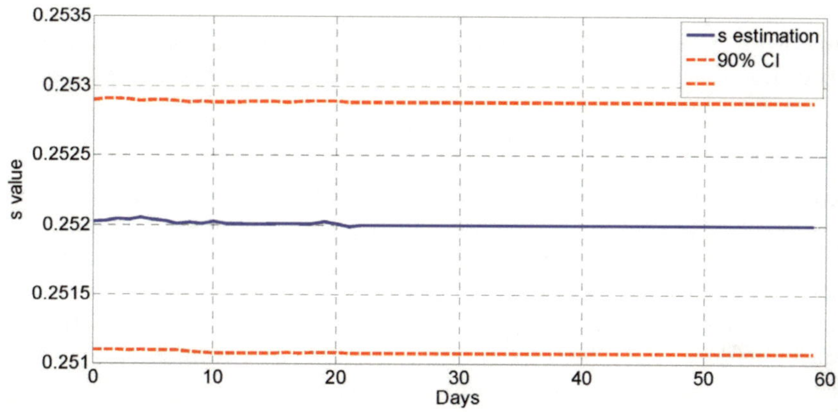

Figure 12.
The s value estimation.

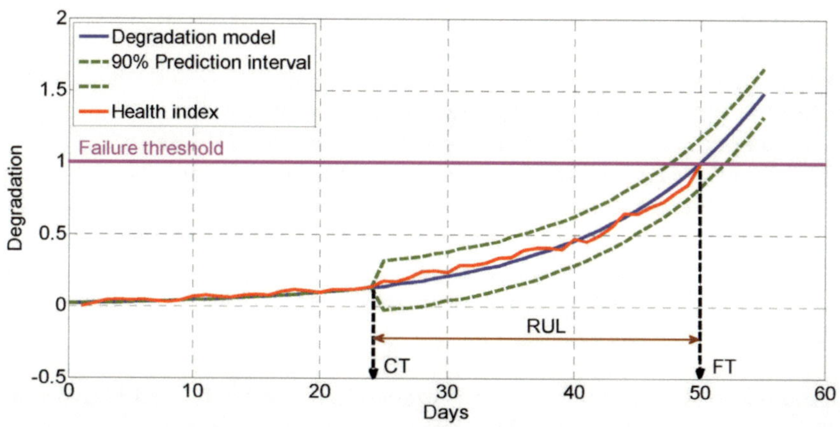

Figure 13.
Particle filter RUL prediction.

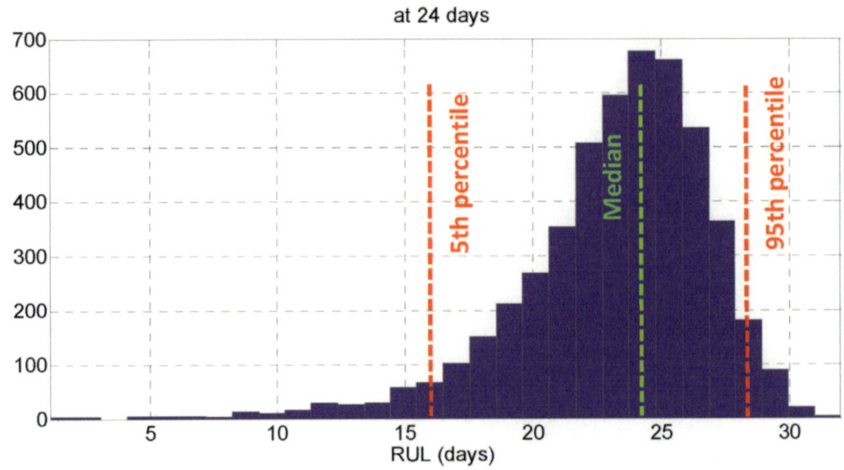

Figure 14.
RUl destribution with percentiles.

threshold time (50 days). The threshold time is given by the last value of HI which is 1. According the **Figure 13**, and (Eq. (16)) the true RUL is 26 days.

$$True_{RUL} = t_{FT} - t_{CT} \tag{16}$$

Figure 14 shows, 5 percentile equal to 16, median equal to 24 and 95 percentile equal to 28 which are caused by 90% of significance level interval. The median value result can be compared with the true RUL in **Figure 13** which is computed using the (Eq. (16)). Therefore, the median value of RUL prediction of 24 days is fairly accurate compared with the true RUL of 26 days. The RUL prediction can be more exact by decreasing the time interval after the current time.

B. Particle filter perfermance

In this section, the discussion is going to compare the robustness and performance of the proposed particle filter approach. Defining prognostics metrics allows comparison and evaluation of different RUL algorithm. By comparing the true and predicted RUL, we can define statistical metrics to measure performance. This comparison needs to be evaluated by the following metrics.

- ***Prognostic horizon (PH)***: The prognostic horizon [24] is described as the difference between the first time when the predicted RUL continuously resides in the accuracy zone and the last cycle time (i.e. End Of Life (EOL)). According the **Figure 13** the accuracy zone has an invariable bound with a value of ±5% error with respect to the last cycle, plotted as two parallel broken lines. The first time when predicted RUL resides in the accuracy zone is at the cycle 24 and the last cycle is 50, thereby PH is 26. The prognostics approaches with a larger PH designate a greater performance, which gives earlier RUL prediction with more reliability.

- ***Convergence***: Eventually, the convergence [25] can be defined with a non-negative error metric (Eq. (17)) of precision or prediction accuracy between the true RUL and predicted RUL. We consider the actual or linear RUL values (denoted X), the predicted RUL (denoted $\check{X}$), and the length of predicted data is n. The RMSE is defined as follows: In this chapter, the relative error is illustrated below:

$$E = |X_i - \hat{X}_i| \tag{17}$$

- ***RUL error***: presents the value computed between true RUL and predicted RUL for each day. The lower value confirms that this method has a good way to predict bearing degraded mode, mathematically the error percent defined as [12, 26].

$$Error(\%) = \frac{X_i - \hat{X}_i}{X_i} \times 100 \tag{18}$$

- ***Root Mean Square Error (RMSE)*** is used to measure the precision of prediction because it is able to make the error and predicted value at the same magnitude [27] We consider the actual or linear RUL values (denoted X), the predicted RUL (denoted $\hat{X}$), and the length of predicted data is n. The RMSE is defined as follows

$$RMSE = \sqrt{\frac{1}{n}\sum_{i=1}^{n}\left(X_i - \widehat{X}_i\right)^2} \tag{19}$$

- ***Mean Absolute Percentage Error (MAPE)*** is generally applied to determine the error size as a percentage. However, it is not advisable with small data sets. The MAPE is defined as follows

$$MAPE = \frac{100}{n}\sum_{i=n}^{n}\left|\frac{X_i - \widehat{X}_i}{X_i}\right| \tag{20}$$

- ***Fitness degree*** (sometimes called the R^2 coefficient), gives an indication of good prognostics when the R^2 value is close to 1. R^2 is defined as fellow

$$R^2 = 1 - \left(\frac{(X_i - \hat{X}_i)^2}{X_i^2}\right) \tag{21}$$

- ***Relative Error Analysis (REA)*** is used to measure the precision which is defined in percentage. It presents relative information between the measurement and the size of data measured. The error is proportional to the size of the RUL being measured. REA is defined as follows

$$REA = \frac{100}{n}\sum_{i=1}^{n}\frac{X_i - \hat{X}_i}{X_i} \tag{22}$$

- ***Accuracy metric (A)*** calculates the "exactitude" between the true RUL and the predicted RUL.. The result of this metric is presented as a percentage, if the accuracy is close to 100% that prove the predicted RUL is similar to the true RUL. The accuracy is defined as

$$A(\%) = 100 \times \left[1 - \frac{|X_i - \hat{X}_i|}{X_i}\right] \tag{23}$$

These above mentioned metrics are used to evaluate PF results as shown in **Table 4**.

Metrics	Particle Filter
Error %	7.69
RMSE	0.2828
MAPE	0.1538
R^2	0.9941
REA	0.1538
A %	92.30
PH	26

Table 4.
Perfermance of the proposed methods.

The PF method adopted in this chapter is called hybrid prognosis approach. This study is based on the combination of data driven approach and exponential model degradation. This combination makes a very powerful prognostics tool. This idea can be extended to combine parameters model and state prediction. After the comparison and the discussion, it is proved that RUL prediction using particle filter method provides a more accurate PH with a 26 cycles. Although we used in the training step only 48% of measured data. It should be noted, this method is able to be applied on any HSSB in wind farm, with an initial value parameter setting (eg; failure threshold).

5. Discussion and comparison with some previous works

Concerning resent work and according to our bibliographic study, this chapter presents a pedagogic implementation of a hybrid prognostic approach based on particle filter for bearing PHM. We encourage all researchers to work on this to have universal approach for bearing prognosis and complete one of the aims of industry 4.0 challenges. In [11], authors have applied the same data of this chapter the RUL prediction based on SVR. A smaller estimation error was found in 60% of training data compared to using 40% of training data. The SVR process is considered as internal RUL prediction. In addition, the SVR model parameters prediction was done after reaching 60% of degradation and that cannot be done online. It is impossible to define online the time where the degradation reaches 60%. In addition, it is hard to build the SVR model and validate it before the recording of the next raw vibration data. Some specific systems need to generate RUL prediction in little times due to the short lifetime of the used bearing and the required precision and excellence such as in robotics or nuclear application. In [5] the proposed Elman Neural Network (ENN) is motivated by a feature extraction from raw vibration data. The feature reduction is considered very important, as non-informative feature will be then discarded. Therefore, the online computational time will be reduced and ENN converge can be easily reached. Consequently, selecting suitable features is a prerequisite for accurate prognostics. ENN based prognosis is powerful but the implementation is costly and complicated. PF used in this chapter is a powerful tool for bearing failure prognosis; the implementation is very easy for and do not require a large data. The main thing to have the best RUL prediction is to build the right exponential degradation model with the true initial parameters, which can make an online PHM. As shown in **Figure 13** the exponential curve trend the HI over 50 days with imperceptible fluctuations see **Table 4**.

6. Conclusion

This chapter introduces how to use approach based on condition monitoring data on HSSB for WTG. Classical statistical features and features derived from SK were used to elicit the bearing health state from the raw vibration data. Trendability, Monotonicity, Prognosability and Suitability are used as metric indices to obtain the corresponding feature for training step. The selected feature is used as Health index for the two proposed prognosis approach in order to predict the best RUL with higher performances.

The acquired results indicate that the Particle filter is more feasible tool for HSSB RUL prediction where the error equal to 7.69% and the degradation model with estimated parameters presents better trends for HI, compared to existing works.

As future work, the proposed method needs to be evaluated in a large amount of HSSB over a very long period. Also, the investigation of time-frequency-domain features will be considered. In addition, we invite next work to focus on external prognostic and to investigate new methodology or adopt some existing ones for dynamic feature selection. This ensures more alignment with industry 4.0 requirements.

Acknowledgements

The authors of this chapter would like to thank green power monitoring systems (GPMS) in USA for the permission to use their bearing data.

Author details

Sharaf Eddine Kramti[1*], Jaouher Ben Ali[1,2], Hugo Andre[3], Eric Brhhoefer[4] and Mounir Sayadi[1]

1 ENSIT, Laboratory of Signal Image and Energy Mastery (SIME), University of Tunis, Tunis, Tunisia

2 ESSTHS- Department of Electronics and Computer Engineering, University of Sousse, Sousse, Tunisia

3 UJM Saint Etienne, LASPI, IUT de Roanne, University of Lyon, France

4 Green Power Monitoring Systems, LLC, Essex, VT, USA

*Address all correspondence to: sharaf.eddine.kramti02@gmail.com

References

[1] Uckun S, Goebel K, Lucas PJ. Standardizing research methods for prognostics. In *2008 International Conference on Prognostics and Health Management IEEE*. 2008:1-10.

[2] Kim, NH, An D, Choi JH. *Prognostics and health management of engineering systems: An introduction*. springer. 2016.

[3] Brahimi M, Medjaher K, Leouatni M, Zerhouni N. Development of a prognostics and health management system for the railway infrastructure—Review and methodology. In *2016 Prognostics and System Health Management Conference (PHM-Chengdu)* IEEE. 2016: 1-8.

[4] Lee J, Wu F, Zhao W, Ghaffari M, Liao L, Siegel D. Prognostics and health management design for rotary machinery systems—Reviews, methodology and applications. *Mechanical systems and signal processing*. 2014; *42*(1-2): 314-334.

[5] Kramti S E, Ben Ali J, Saidi L, Sayadi M, Bouchouicha M, Bechhoefer E. A neural network approach for improved prognostics of wind turbine generators. The European Physical Journal Applied Physics. 2021

[6] Atamuradov V, Medjaher K, Dersin P, Lamoureux B, Zerhouni N. Prognostics and health management for maintenance practitioners-review, implementation and tools evaluation. *International Journal of Prognostics and Health Management*. 2017 ;*8*(060); 1-31.

[7] Camci, F., Medjaher, K., Zerhouni, N. and Nectoux, P. (2013), Feature Evaluation for Effective Bearing Prognostics. Qual. Reliab. Engng. Int., 29: 477-486.

[8] Jardine AK, Lin D, Banjevic D. A review on machinery diagnostics and prognostics implementing condition-based maintenance. *Mechanical systems and signal processing*. 2006; *20*(7): 1483-1510.

[9] Saidi L, Ben Ali J, Benbouzid M, Bechhofer E. An integrated wind turbine failures prognostic approach implementing Kalman smoother with confidence bounds. *Applied Acoustics*. 2018; *138:* 199-208.

[10] Kramti SE, Ben Ali J, Saidi L, Sayadi M, Bechhoefer E. Direct wind turbine drivetrain prognosis approach using Elman neural network. In *2018 5th International Conference on Control, Decision and Information Technologies (CoDIT)* IEEE. 2018: 859-864.

[11] Saidi L, Ben Ali J, Bechhoefer E, Benbouzid M. Wind turbine high-speed shaft bearings health prognosis through a spectral Kurtosis-derived indices and SVR. *Applied Acoustics*. 2017 ;*120:* 1-8.

[12] Elforjani M, Shanbr S. Prognosis of bearing acoustic emission signals using supervised machine learning. *IEEE Transactions on industrial electronics.* 2017 ;*65*(7): 5864-5871.

[13] SKF bearing catalog. Available at: https://www.bearing-king.co.uk/bearing/32222-j2-skf-metric-single-row-taper-roller-bearing/2974 [accessed date: 05-01-2019].

[14] Saini MK, Aggarwal A. (2018). Detection and diagnosis of induction motor bearing faults using multiwavelet transform and naive Bayes classifier. *International Transactions on Electrical Energy Systems*. 2018; *28*(8), e2577.

[15] Antoni J, Randall RB. The spectral kurtosis: application to the vibratory surveillance and diagnostics of rotating machines. *Mechanical systems and signal processing.* 2006 ;*20*(2): 308-331.

[16] Antoni J. The spectral kurtosis: a useful tool for characterising

non-stationary signals. *Mechanical systems and signal processing*. 2006; *20*(2): 282-307.

[17] Kramti SE, Saidi L, Ben Ali J, Sayadi M, Bechhoefer E. Particle Filter Based Approach for Wind Turbine High-Speed Shaft Bearing Health Prognosis. In *2019 International Conference on Signal, Control and Communication (SCC 2019)* IEEE. 2019.

[18] E, Sutrismo. H, Oh. A, Vasan. M, Pecht." Estimation of Remaining Useful Life of Ball Bearings using Data Driven Methodologies" *Prognostics and Health Management (PHM)*, IEEE, 2012.

[19] D. An, J-Ho. Choi, N. H. Kim. Prognostics 101: A tutorial for particle filter-based Prognostics algorithm using Matlab. *Reliability Engeneering and System Safety.* 2013; 115:161-169.

[20] N. H. Kim, D. An, J-H. Choi, "Prognostics and Health Management of Engineering Systems: An Introduction," *Springer International Publishing Switzerland* 2017.

[21] M. Jouin, R. Gouriveau, D. Hissel,M-C. Péra, N. Zerhouni, "Particle filter-based prognostics: Review, discussion and perspectives," Mechanical Systems and Signal Processing vol. 72-73, pp. 2-31, 2016.

[22] Javed K, Gouriveau R, Zerhouni N, Nectoux P. A feature extraction procedure based on trigonometric functions and cumulative descriptors to enhance prognostics modeling. In *2013 IEEE Conference on Prognostics and Health Management (PHM).2013:* 1-7.

[23] Javed K. *A robust & reliable Data-driven prognostics approach based on extreme learning machine and fuzzy clustering* (Doctoral dissertation) 2014.

[24] Saxena A, Celaya J, Saha B. On applying the prognostic performance metrics. *Annual conference of the prognostics and health management society*. 2009.

[25] A. Saxena, J. Celaya, B. Saha, S. Saha and K. Goebel, "Evaluating algorithm performance metrics tailored for prognostics," *2009 IEEE Aerospace conference*, Big Sky, MT, 2009, pp. 1-13.

[26] Sutrisno E, Oh H, Vasan ASS, Pecht M. Estimation of remaining useful life of ball bearings using data driven methodologies. In *2012 IEEE Conference on Prognostics and Health Management*. 2012: 1-7.

[27] Ben Ali J, Hamdi T, Fnaiech N, Di Costanzo V, Fnaiech F, Ginoux JM. Continuous blood glucose level prediction of Type 1 Diabetes based on Artificial Neural Network. *Biocybernetics and Biomedical Engineering*. 2018 ;*38*(4): 828-840.

Section 3

Modelling and Optimization

Chapter 4

Use of Discrete-Time Forecast Modeling to Enhance Feedback Control and Physically Unrealizable Feedforward Control with Applications

Derrick K. Rollins

Abstract

When the manipulated variable (MV) has significantly large time delay in changing the control variable (CV), use of the currently measured CV in the feedback error can result in very deficient feedback control (FBC). However, control strategies that use forecast modeling to estimate future CV values and use them in the feedback error have the potential to control as well as a feedback controller with no MV deadtime using the measured value of CV. This work evaluates and compares FBC algorithms using discrete-time forecast modeling when MV has a large deadtime. When a feedforward control (FFC) law results in a physically unrealizable (PU) controller, the common approach is to use approximations to obtain a physically realizable feedforward controller. Using a discrete-time forecast modeling method, this work demonstrates an effective approach for PU FFC. The Smith Predictor is a popular control strategy when CV has measurement deadtime but not MV deadtime. The work demonstrates equivalency of this discrete-time forecast modeling approach to the Smith Predictor FBC approach. Thus, this work demonstrates effectiveness of the discrete-time forecast modeling approach for FBC with MV or DV deadtime and PU FFC.

Keywords: Model Predictive Control, Nonlinear Dynamic Modeling, Artificial Pancreas

1. Introduction

Modeling data is critical to the advancement of information and data science on many levels and in many areas. Accurately modeling data is often important to system monitoring, understanding, and control; and thus, ultimately to the advancement of technology.

A characteristic of data that that is not well understood, even by those in the physical sciences, is dynamic behavior. However, the behavior of Covid-19, which is inherently dynamic, has forced wide-spread conversations from even the non-science community about such terms as lag and deadtime. Just as understanding the attributes of Covid-19 dynamically can lead to intelligent and thus, safe behavior, good decision making and preparation, and even save lives, the lack of understanding can do the opposite. Thus, the better our understanding of dynamic behavior is, physically and biologically, the better our understanding of data will be, leading only to better solutions to many problems facing society.

Forecast modeling is a type of predictive modeling that uses current and anticipated future input values to predict values for outputs in the future. For example, forecast modeling is used to predict the wind velocity in a certain region five days into the future. Another example is forecast modeling to predict the number of deaths caused by a virus a week into the future. This chapter focuses on the application of forecast modeling to enhance feedback control (FBC) and feedforward control (FFC) when the problem is physically unrealizable (PU).

A PU system is a mathematical phenomenon of a dynamic system. It occurs in two ways. The first one is when the order of the differential equation for the output is less than the order of the differential equation for the input. An example is the development of a FFC law determined from a load transfer function divided by the process transfer of a lower order. The second one is when an output depends on deadtime that has a negative value, in effect causing a dependence on future values of a time dependent variable(s). This also occurs in FFC when the load transfer function has a smaller deadtime than the process transfer function. The most common approach for addressing a PU system is to use approximation(s) to make it physically realizable. However, such approximations can lead to large modeling errors, thus leading to unacceptable control.

The dynamic modeling literature [1, 2] defines ***causality*** somewhat differently than the statistics literature. More specifically, "if a system['s] output depends on the future input values ... the system is noncausal [2]." **This definition is synonymous with PU**, it seems. In forecast modeling, all values of inputs are before the forecast distance in the future. They can be in the future, but not a distance beyond the forecast time. Another description for PU in the dynamic modeling literature is *improper transfer function*.

In FFC, MV is the output and depends on input changes. When MV has a deadtime of θ_{MV}, for example, it takes this time before a change in MV affects CV. Within this period, other inputs may change that impact CV before the change in MV does. An example is the control of blood glucose concentration (BGC) in type 1 diabetes. The deadtime for insulin infusion is much larger than the deadtime for carbohydrate consumption. The best approach for control of BGC is to consider the timing and the amount of carbohydrates consumed and to bolus this with a determined amount of insulin, a calculated amount of the time before the meal. This procedure is just a manual type of PU FFC practiced by people with type 1 diabetes.

The use of causality in the statistics literature seeks to distinguish it from correlation. Thus, in the statistics literature, causality is not focused exclusively on dynamic systems (e.g., only those with lag or deadtime) but a ***cause-and-effect*** relationship between input and output, that can be nondynamic [3]. For forecast modeling, ***cause-and-effect*** is not essential if the model is accurate. However, in control, ***cause-and-effect*** modeling is essential. A PU system does not have an exact solution which would be a continuous-time solution. However, a discrete-time solution can be determined directly from the PU continuous-time structure. Hence, this work uses highly structured discrete-time forecasting and FFC models.

1.1 Objective and contribution of the work

Moreover, the primary objective of this chapter is to apply a novel discrete-time forecasting modeling methodology to systems with large MV deadtime in FBC and PU FFC without physical realizable approximations. For this scope, FBC is examined and evaluated under three prediction horizons: 1. None – Classical FBC that uses the currently measured value of CV [4, 5]; 2. θ_{MV} – Feedback Predictive Control (FBC) [6] and; greater than θ_{MV} – Model Predictive Control (MPC) [7, 8]. The Smith Predictor (SP) [9] is a novel FBC approach when θ_{MV} = 0 and there is deadtime in the measurement of CV. This work shows that FBPC gives equivalent control of the SP, but also has the advantage that it is applicable when $\theta_{MV} > 0$, which the SP is not. In addition, this work reveals the detrimental use of the bias correction as given in the block diagram of the SP and used widely in process control [4]. Thus, this work proposes a better bias correction method. Finally, this works presents a novel discrete-time PU FFC algorithm that is multiple-input and single-output and hence, is able to treat complex multiple-input feedforward model structures. Although MPC is a FBC approach, comparison is made to illustrate the potential improvement of FFPC over model-based predictive FBC.

2. Physically unrealizable

Physical unrealizability (PU) is an anomaly that is strictly an artifact of a dynamic system. A dynamic system has at least one process state (i.e., output or response) that does not change to its new value immediately when input changes occur that cause its value (i.e., level) to change. This behavior contrasts with a nondynamic system that changes to its new state immediately when inputs change (also, called "disturbances").

There are two basic dynamic phenomena – time lag and time delay (also called "dead time") which are shown in **Figure 1**. This figure illustrates nondynamic and dynamic relationships for the response, y, to a step change in the input, x, occurring

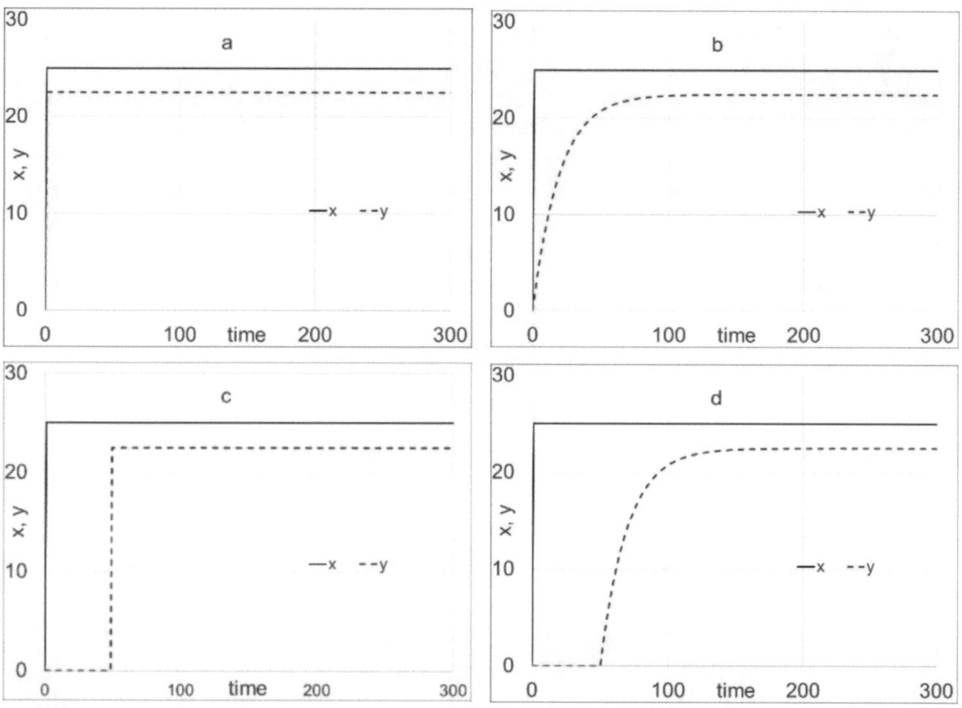

Figure 1.
Response y *to a step change in* x*: a. nondynamic; b. lag; c. time delay and; d. lag and time delay.*

at time (t) = 0. As shown, for the nondynamic response (a), y changes immediately to its new steady state value. Lag (b) is shown by the change in y starting to occur at t = 0, but monotonically increasing over time to its new steady state value. Time delay (c) is shown by the change in y to its new value occurring θ time later. Lag and time delay (d) are shown by y starting to change θ time later and then monotonically increasing over time to its new steady state value. "Everyday" examples of nondynamic changes are eyes opening (x) and immediate sight (y) and turning on the radio in a car (x) and hearing it (y) immediately. A dynamic time delay example is lightening occurring very far away. It occurs when one sees the lightening, but the thunder is delayed and occurs at a significant time after seeing the lightening. It does not build up to its final value, there is just a big boom that occurs, essentially, at once. A dynamic change with lag occurs when a person has been out in the cold for a while and their skin temperature is quite cold and when they move to a warmer environment, their temperature starts to rise but it takes time for the temperature to reach its new level in this warmer environment.

When a system is dynamic, its mass and/or internal energy changes over time, being driven to a new state due to input changes, arriving there at a time different than when the input was changed. Mathematically (and theoretically) this is seen as a time-order differential equation. Such an equation is given in terms of input x and output y in Eq. (1).

$$\begin{aligned} &a_n\frac{d^n y(t)}{dt^n}+a_{n-1}\frac{d^{n-1}y(t)}{dt^{n-1}}+\cdots+a_1\frac{dy(t)}{dt}+a_0y(t)= \\ &b_m\frac{d^m x(t-\theta)}{dt^m}+b_{m-1}\frac{d^{m-1}x(t-\theta)}{dt^{m-1}}+\cdots+b_1\frac{dx(t-\theta)}{dt}+b_0x(t-\theta) \end{aligned} \tag{1}$$

The dynamic form represented by Eq. (1) is PU if $n < m$, if θ is negative, or if both are true. More specifically, the output, y, which depends on the input, x, cannot have a time dependent derivative structure that is of lower order than the variable that causes it to change. The response of a system to a disturbance also cannot have negative time delay. A system cannot respond to a disturbance before it occurs. There is no true solution for these conditions.

However, there are ways to address these PU cases in practice. For the first case, $n < m$, discrete-time backwards different derivatives can be used to approximate the continuous-time derivatives. This approach should provide adequate accuracy when the sampling time is constant and sufficiently small, and sensor noise is not too great. Eq. (2) illustrates this approximation when m = 2, n = 1, a constant sampling time, Δt, and with θ = 0 (for simplicity). Note, since $y(t)$ cannot be immediately affected by $x(t)$, $x_{t\text{-}\Delta t}$ is used to approximate $x(t)$.

$$\begin{aligned} &a_1\frac{dy(t)}{dt^n}+a_0y(t)=b_2\frac{d^2x(t)}{dt^2}+b_1\frac{dx(t)}{dt}+b_0x(t) \\ &\Rightarrow y_t\approx\frac{y_{t-\Delta t}+\left(\frac{b_2}{\Delta t}+b_1+b_0\Delta t\right)x_{t-\Delta t}-\left(\frac{2b_2}{\Delta t}+b_1\right)x_{t-2\Delta t}+\frac{b_2}{\Delta t}x_{t-3\Delta t}}{1+a_0\Delta t} \end{aligned} \tag{2}$$

Thus, digital and sensor technologies, among other advancements, have significantly contributed to an ability to approximate Eq. (1) when $n < m$.

For the other case, i.e., when the time delay is a negative value such as a -5Δt (e.g., in FFC when deadtime for the disturbance variable (θ_{DV}) is smaller than the deadtime for the manipulated variable (θ_{MV})), Eq. (2) becomes

$$a_1\frac{dy(t)}{dt^n}+a_0y(t)=b_2\frac{d^2x(t+5\Delta t)}{dt^2}+b_1\frac{dx(t+5\Delta t)}{dt}+b_0x(t+5\Delta t)$$

$$\Rightarrow y_t\approx\frac{y_{t-\Delta t}+\left(\frac{b_2}{\Delta t}+b_1+b_0\Delta t\right)x_{t+5\Delta t-\Delta t}-\left(\frac{2b_2}{\Delta t}+b_1\right)x_{t+5\Delta t-2\Delta t}+\frac{b_2}{\Delta t}x_{t+5\Delta t-3\Delta t}}{1+a_0\Delta t}$$

$$=\frac{y_{t-\Delta t}+\left(\frac{b_2}{\Delta t}+b_1+b_0\Delta t\right)x_{t+4\Delta t}-\left(\frac{2b_2}{\Delta t}+b_1\right)x_{t+3\Delta t}+\frac{b_2}{\Delta t}x_{t+2\Delta t}}{1+a_0\Delta t} \tag{3}$$

Thus, as shown by Eq. (3), the output depends on future values of the input. However, discrete-time modeling provides a means to express an approximate solution in a PU form where knowledge of future changes allows approximation of the output. An example where this type of approximation is applied, is the control of blood glucose concentration (BGC) for people with type 1 diabetes [10]. The manipulated variable (MV) for the automatic regulation of the exogenous insulin infusion from a servo-mechanical pump can have a deadtime of 60 minutes and carbohydrates from meals can have a deadtime of 30 minutes [11], resulting in $x(t)$ becoming $x(t + 30)$ in the numerator of the FFC law, with MV as the output variable. Moreover, for these values, a change in insulin flow rate will take one hour to begin to lower BGC. During this period, eating can increase BGC. People with type 1 diabetes understand this phenomenon and will bolus their insulin infusion, based on when they will eat and how many carbohydrates they will eat. This is called "a meal announcement" [12, 13]. But just as this idea is applied to carbohydrates, it can be applied to other variables with dead times less than MV such as stress, exercise, etc. As one can imagine, the relationships of such a set of variables on BGC is quite complex and accurately modeling their relationships and automatic feedback/feedforward control (FBFFC), with accurate announcements, appears to be the most viable one for success. In the content to follow, we focus on dynamic modeling with application to feedback control (FBC) and FFC when time delay in MV is significantly larger than time delay in disturbances. For this situation, when CV is the output, forecast modeling is necessary and when MV is the output, future announcement (i.e., knowledge) is needed for any variable with a dead time less than that of MV.

3. Discrete-time forecast dynamic modeling

Accurate forecast dynamic modeling in the context of process control has two critical applications. One is accurately *forecasting* CV at least θ_{MV} distance into the future, depending on the type of model-based control algorithm being used in FBC. The other one is an accurate *cause-and-effect* model for CV that is inverted for determining MV as a function of disturbances in FFC. Empirical modeling methods (EMM) (i.e., the so-called "data-driven" methods such as linear regression and artificial neural networks) are fit to a correlation structure and should not be used for *cause-and-effect* modeling unless the modeling data are generated from a statistical experimental design covering the full range of the operating (i.e., input) space. This input space will be orthogonal and prevent extrapolation, which is risky for EM. For "freely existing data" or any data not generated from a statistical experimental design, accurate EM for CV forecasting is possible. However, when modeling data are not generated by a statistical experimental design, it would not be wise to use EM when *cause-and-effect* modeling is needed since EMM are data driven and not knowledge driven, rely on high levels of parametrization, do not have structures or parameters that are physically interpretable based on first principles modeling, and are typically very risky for, even slight, extrapolation. In contrast, first principles model structures are: 1. nonlinear and thus, naturally break down correlation structures in the input data; 2. have physically interpretable parameters; and 3. often physical constraints with a theoretical basis. Nonetheless, theoretically based modeling of real data outside a controlled environment such as a lab, is often some combination of empiricism and first principles knowledge, which is essentially a "hybrid model" that is often called "gray box" or "semi-empirical" models. Models that are fully theoretical in derivation and structure but use data to obtain unknown physically interpretable model parameters, are classified in this document as semi-theoretical models.

Theoretically structured dynamic systems can be linearized (i.e., approximated) in time dependent variables (i.e., $x = x(t)$) while maintaining their time derivative structures (i.e., the order of derivatives will remain intact) and physical parametrization. For example, Eq. (4) represents the result of a dynamic overall mass balance on a process tank with one inlet stream with flow rate, $q_1(t)$, and one outlet stream though a hand valve with flow rate, $q(t) = h^2(t)/\mathrm{R_v}$, where h is the tank level, R_v is the resistance to flow through the valve, and A is the cross-sectional area of the tank. The density and temperature of the fluid in the tank is constant in this example. Using a 1st order Taylor Series approximation to linearize all time dependent variables in Eq. (4), gives the solution in Eq. (5), where the "′" represents a variable as a deviation from its initial steady state at $t = 0$.

$$A\frac{dh(t)}{dt} = q_1(t) - R_V^{-1}h^2(t) \tag{4}$$

$$A\frac{dh'(t)}{dt} = q_1'(t) - 2R_V^{-1}h(0)h'(t) \tag{5}$$

Rearranging Eq. (5) into the form of Eq. (1) gives:

$$\begin{aligned} a_1\frac{dh'(t)}{dt} + a_0h'(t) = b_0q_1'(t) \\ \tau\frac{dh'(t)}{dt} + h'(t) = Kq_1'(t) \end{aligned} \tag{6}$$

where $a_1 = A\left[2R_V^{-1}h(0)\right]^{-1} = \tau,\ a_0 = 1\ and\ b_0 = K = \left[2R_V^{-1}h(0)\right]^{-1}$. Eq. (6) is a first-order dynamic relationship with time constant, τ, and steady-state gain, K, and is represented in "standard form [4, 5]." Many dynamic processes can be approximated accurately by either a first-order-plus-deadtime (FOPDT) or second-order-plus-deadtime (SOPDT) structure [4, 5].

Eq. (7) gives a second-order version of Eq. (1) for inputs $x_i',\ i = 1,\ \ldots, p$, and unity gain, with y replaced by v_i'. Thus, Eq. (7) represents the dynamic response due to x_i', in the units of x_i'. Eq. (8) represent the dynamic response of y as a function of $v_i''s$ where $f(V)$ is an unrestricted mathematical function that maps each v_i' to the units of the output variable in standard form. Thus, it is $f(V)$ that transforms the linear dynamic inputs into the nonlinear dynamic and static response for the output y.

$$\tau_i^2\frac{d^2v_i(t)}{dt^2} + 2\tau_i\zeta_i\frac{dv_i(t)}{dt} + v_i(t) = \tau_{ai}\frac{dx_i(t-\theta_i)}{dt} + x_i(t-\theta_i) \tag{7}$$

$$y(t) = f(\mathrm{V}(t)) + \varepsilon(t) = \eta(t) + \varepsilon(t) \tag{8}$$

where $\mathrm{V}(t)$ is a vector of the $v_i''s$ and the estimate of $y(t)$, denoted as $\hat{y}(t)$, is equal to the estimate of $\eta(t)$, $\hat{\eta}(t)$. This hybrid dynamic modeling structure is called a Wiener network [14] and is in a class of structures that are called "block-oriented models." The block diagram for this network is shown in **Figure 2**.

Rollins, et al. [15] developed a multiple-input, single-output, discrete-time, non-linear Wiener dynamic approach using backwards difference derivatives based on Eqs. (7) and (8). Using a backward difference approximation applied to a sampling interval of Δt, an approximate discrete-time form of Eq. (7) is obtained (for p inputs):

$$v_{i,t} = \delta_{1,i}v_{i,t-\Delta t} + \delta_{2,i}v_{i,t-2\Delta t} + \omega_{1,i}x_{i,t-\theta_i-\Delta t} + \omega_{2,i}x_{i,t-\theta_i-2\Delta t} \tag{9}$$

where $\omega_{2,i} = 1 - \delta_{1,i} - \delta_{2,i} - \omega_{1,i}$ to satisfy the unity gain constraint with

$$\delta_{1,i} = \frac{2\tau_i^2 + 2\tau_i\zeta_i\Delta t}{\tau_i^2 + 2\tau_i\zeta_i\Delta t + \Delta t^2} \tag{10}$$

$$\delta_{2,i} = \frac{-\tau_i^2}{\tau_i^2 + 2\tau_i\zeta_i\Delta t + \Delta t^2} \tag{11}$$

$$\omega_{1,i} = \frac{(\tau_{ai} + \Delta t)\Delta t}{\tau_i^2 + 2\tau_i\zeta_i\Delta t + \Delta t^2} \tag{12}$$

After obtaining $v_{i,t}$ for each input i (i = p is MV), the modeled output value is determined by substituting these results into f ($\boldsymbol{V}$), such as

$$\begin{aligned}\eta_t = f(\mathrm{V}) = a_0 + a_1v_{1,t} + \cdots + a_pv_{p,t} + b_1v_{1,t}^2 + \cdots + b_pv_{p,t}^2 \\ +c_{1,2}v_{1,t}v_{2,t} + \cdots + c_{p-1,p}v_{p-1,t}v_{p,t}\end{aligned} \tag{13}$$

Modification of Eq. (13) for forecasting $\eta(t)$ a distance θ_{MV} into the future with p = 3, for example, gives

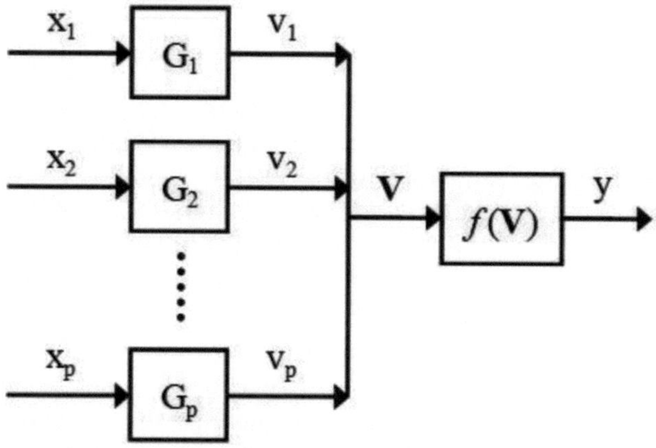

Figure 2.
Block diagram for the wiener network with p *inputs and one output. Each input,* x_i, *is passed through their own unity gain linear dynamic block,* G_i, *after which these unobservable intermediate outputs are collected and passed through a single unrestricted static gain function,* f(**V**), *to produce the output,* y.

$$\eta_{t+\theta_{MV}} = a_0 + a_1 v_{1,t+\theta_{MV}} + \cdots + a_3 v_{3,t+\theta_{MV}} + b_1 v_{1,t+\theta_{MV}}^2 + \cdots + b_3 v_{3,t+\theta_{MV}}^2 + c_{1,2} v_{1,t+\theta_{MV}} v_{2,t+\theta_{MV}} + \cdots + c_{2,3} v_{2,t+\theta_{MV}} v_{3,t+\theta_{MV}} \quad (14)$$

where a_i, b_i, and $c_{i,j}$, denote the linear, quadratic and interaction parameters for i = 1, 2, 3 and j = 2 and 3, and

$$v_{1,t+\theta_{MV}} = \delta_{1,1} v_{1,t+\theta_{MV}-\Delta t} + \delta_{2,1} v_{1,t+\theta_{MV}-2\Delta t} + \omega_{1,1} x_{1,t+\theta_{MV}-\theta_{DV_1}-\Delta t} + \omega_{2,1} x_{1,t+\theta_{MV}-\theta_{DV_1}-2\Delta t} \quad (15)$$

$$v_{2,t+\theta_{MV}} = \delta_{1,2} v_{2,\theta_{MV}-\Delta t} + \delta_{2,2} v_{2,t+\theta_{MV}-2\Delta t} + \omega_{1,2} x_{2,t+\theta_{MV}-\theta_{DV_2}-\Delta t} + \omega_{2,2} x_{2,t+\theta_{MV}-\theta_{DV2}-2\Delta t} \quad (16)$$

$$\begin{aligned} v_{3,t+\theta_{MV}} &= \delta_{1,3} v_{3,t+\theta_{MV}-\Delta t} + \delta_{2,3} v_{3,t+\theta_{MV}-2\Delta t} \\ &\quad + \omega_{1,3} x_{3,t+\theta_{MV}-\theta_{MV}-\Delta t} + \omega_{2,3} x_{3,t+\theta_{MV}-\theta_{MV}-2\Delta t} \\ &= \delta_{1,3} v_{3,t+\theta_{MV}-\Delta t} + \delta_{2,3} v_{3,t+\theta_{MV}-2\Delta t} + \omega_{1,3} x_{3,t-\Delta t} + \omega_{2,3} x_{3,t-2\Delta t} \end{aligned} \quad (17)$$

where the θ's are integer multiples of Δt. Depending on the rate of change of CV, forecast accuracy (and hence, control) can suffer significantly by setting θ_{MV} - θ_{DV} to zero. Developers of BGC devices that use current sensor glucose measurements in the feedback error restrict these devices for use only during long sleeping periods when BGC changes very slowly. For an application such as automatic BGC control, with a very large deadtime for MV and many disturbances with smaller deadtime than MV, that are nonlinear and interactive, the required accuracy for forecasting BGC is quite challenging. However, as health monitoring sensor technology continues to advance, forecast modeling accuracy continues to improve. The strengths of the method presented in this chapter are the use of dynamic structures that are embedded in first principles modeling; that is, they have physically interpretable parameters embedded in highly nonlinear structures (Eqs. (10)–(12)) with physical constraints such as $\omega_{2,i} = 1 - \delta_{1,i} - \delta_{2,i} - \omega_{1,i}$, $\tau_i > 0$ and $\zeta_i > 0$, for all i. While the method has these strengths for forecasting [16, 17], these strengths are quite critical in FFC applications where *cause-and-effect* modeling is essential. Next PU is examined from a control perspective – FBC first and then feedback feedforward (FBFFC).

4. FBC when θ_{MV} is large

As discussed above, a change in MV will not affect CV until θ_{MV} time into the future. When θ_{MV} is 0, the feedback error for FBC is, rightly, $e_t = Y^{set} - y_t$, where y_t is the measured value of CV at the current time, t. When θ_{MV} is not 0, the equivalent feedback error is $e_t = Y^{set} - y_{t+\theta MV}$ which is unknown because $y_{t+\theta MV}$ is not obtained until time t + θ_{MV}. This section describes and compares three FBC approaches when θ_{MV} is not 0. The first one is classical FBC [4, 5] which uses $e_t = Y^{set} - y_t$. The second one is feedback predictive control (FBPC) [6] which uses $e_t = Y^{set} - \hat{y}_{t+\theta_{MV}}$. The third one is model predictive control (MPC) [18–20]. The MPC control law is for CV to be equal to Y^{set}, J time steps after t + θ_{MV} while holding the current value of MV fixed [4]. Thus, J is the only controller tuning parameter for MPC. More specifically, its feedback error is: $e_t = Y^{set} - \hat{y}_{t+\theta_{MV}+J\Delta t}$. Hence, the MPC prediction horizon is longer than FBPC by the amount $J\Delta t$. For MPC, the optimal value of $J\Delta t$ will tend to increase as the time lag of y_t increases for changes in MV. Since forecasting accuracy typically decreases as the distance into the future increases, MPC control can significantly deteriorate as $J\Delta t$ increases.

These three control algorithms were compared by [6] in their ability to automatically control CV for a true FOPDT process with K = 1, τ = 10 and 50 min, θ_{MV} = 3 min, and a sampling time, Δt, equal to 1 min. FBPC and FBC were PI-Controllers with tuning parameters to give the best response with little, to no, overshoot for a unit step change in the set point at time t = 0.

Figure 3 presents the results of this study found in [6]. As shown, CV (y) is on the left and MV (M) is on the right. The top row represents τ = 10 (J = 3, 8, and 20) and the bottom row represents τ = 50 (J = 8, 20, and 30). As shown, as J decreases, y reaches the set point faster and overshoots it for the lowest values of J. FBPC reaches the set point much faster than MPC, even when MPC overshoots the set point. As τ increases, MPC takes longer to reach the set point, but this is not the case for FBPC and FBC. FBC reaches the set point faster without overshooting than MPC for the case with the larger τ. Moreover, depending on J and τ, FBC and MPC can reach the set point about the same time without overshooting the set point. However, FBPC has a faster response and reaches the set point much earlier than FBC and MPC in all cases. MV for FBPC has an initial "kick" much greater than FBC or MPC. However, its MV quickly drops below that of FBC and MPC and has significantly less movement in both cases as shown in **Figure 3**. Thus, as expected, because of the longer control horizon, which increases as τ increases, MPC responded slower than FBPC in reaching and staying at the new set point. Similar conclusions were seen in a comparison of FBPC and MPC in this article [6] for a simulated CSTR. Nonetheless, the main conclusion is that there are model-based forecasting FBC algorithms that are viable alternative to classical FBC when θ_{MV} is appreciably large.

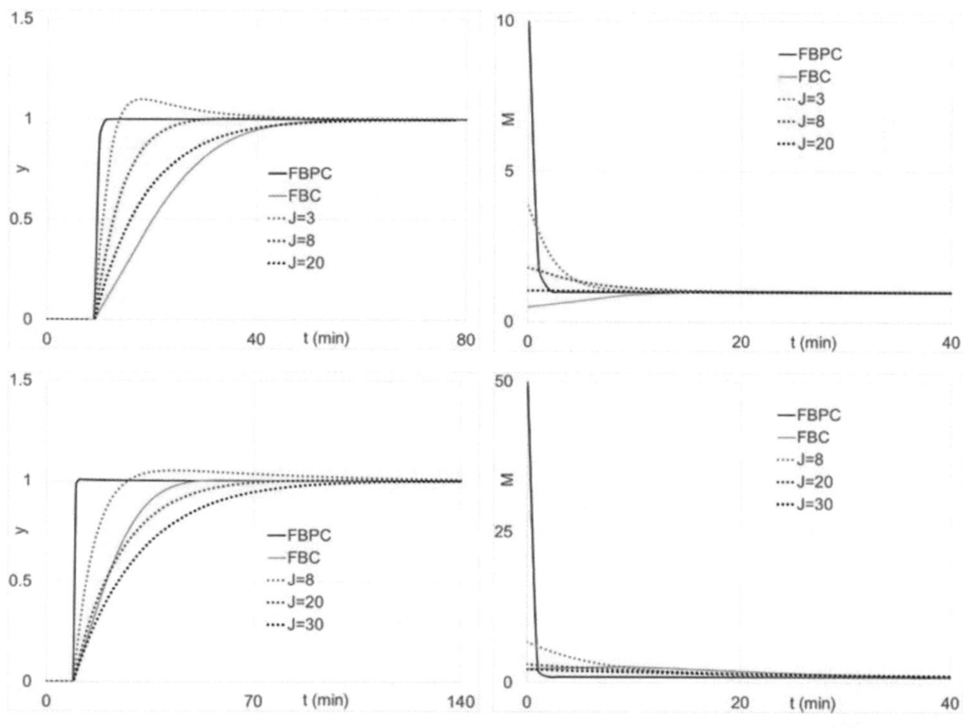

Figure 3.
CV (y) responses (left panels) and MV (M) changes (right panels) for FBPC, FBC and MPC for the FOPDT process. The top case is for τ = 10 (for MPC with J = 3, 8, and 20) and the bottom one is for τ = 50 (for MPC with J = 8, 20, and 30) [6].

5. FFC when θ_{MV} is large

All classical process control textbooks (e.g., [4, 5]) derive the FFC algorithm from a block diagram and give the FFC transfer function for each DV, G_f, as $-G_{DV}/G_{MV}$. The outputs from each G_f are added together to form the multiple-disturbance feedforward control law. An example when both G_{DV} and G_{MV} are FOPDT is:

$$G_f = -\frac{G_{DV}}{G_{MV}} = -\frac{\frac{K_{DV}e^{-\theta_{DV}s}}{\tau_{DV}s+1}}{\frac{K_{MV}e^{-\theta_{MV}s}}{\tau_{MV}s+1}} = -\frac{K_{DV}}{K_{MV}}\frac{\tau_{MV}s+1}{\tau_{DV}s+1}e^{(\theta_{MV}-\theta_{DV})s} = -\frac{K_{DV}}{K_{MV}}\frac{\tau_{MV}s+1}{\tau_{DV}s+1}e^{\Delta\theta s} \tag{18}$$

Thus, G_f will be PU when $\theta_{MV} > \theta_{DV}$, i.e., when $\Delta\theta > 0$. Typically, this limitation is addressed by just setting $\Delta\theta$ to 0 or increasing τ_{MV} to $\tau_{MV} + \Delta\theta$ and setting $e^{\Delta\theta s}$ to 1 [4]. This approximation is usually acceptable when $\Delta\theta$ is small, as commonly found in chemical processes. However, modern applications of process control have gone beyond chemical processes to biological processes where transport is cellular, slow, and complex (i.e., not well understood). A common example is exogenous insulin taken by people with diabetes as mentioned above [21]. Insulin deadtime is significantly greater than the deadtime for carbohydrate intake and other disturbances. Recent advancements in activity trackers measure multiple variables that likely affect BGC, and most, if not, all have a smaller deadtime than insulin [15–17].

Process Control textbooks commonly describe additive and linear dynamic FFC and present the algorithms in the continuous-time Laplace (s-) domain. This section presents a FFC approach that is: 1. given in the time domain; 2. discrete-time; 3. able to treat all types of non-additive behavior as well as nonlinear dynamic and static behavior and; 4. combines all disturbances functionally into one FFC law (i.e., all the DV's enter one FFC equation). A block diagram of this FFC approach based on the Wiener network is given in **Figure 4**. As shown, the modeled disturbances are

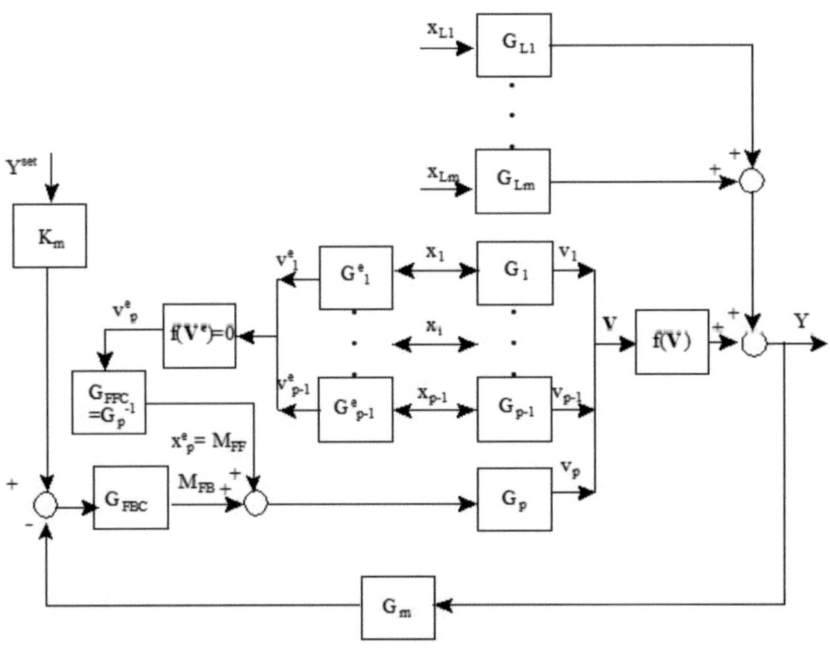

Figure 4.
Multiple-input FBC/FFC block diagram for a p-input wiener network FFC model [22].

x_1 to x_{p-1} and x_p is MV. Inputs x_1 to x_{p-1} pass through their dynamic blocks to produce v_1 to v_{p-1}. The FFC law associated with **Figure 4** is:

$$e_{ffc} = \left(Y^{set} - f(V^e)\right)\big|_{x_p^e} = 0 \tag{19}$$

where $f(V^e)$ is defined in Eq, (8) with the superscript e associating it with the FFC law, i.e., $x_p^e = x_{MV}^e$ = value of MV that makes $e_{ffc} = 0$, and thus, satisfying Eq. (19).

For $p = 2$, i.e., one disturbance and MV, and FOPDT structures for both inputs, and application of linear forms for Eq. (13) (for simplicity) into Eq. (19) gives:

$$\begin{aligned}
e_{ffc} &= \left(Y^{set} - a_0 - a_1 v_{1,t} - a_2 v_{2,t}\right)\big|_{x_2} = Y^{set} - a_0 \\
&\quad - a_1\left[\delta_{1,1} v_{1,t+\theta_{MV}-\Delta t} + \delta_{2,1} v_{1,t+\theta_{MV}-2\Delta t} + \omega_{1,1} x_{1,t+\Delta\theta-\Delta t} + \omega_{2,1} x_{1,t+\Delta\theta-2\Delta t}\right] \\
&\quad - a_2\left[\delta_{1,2} v_{2,t+\theta_{MV}-\Delta t} + \delta_{2,2} v_{2,t+\theta_{MV}-2\Delta t} + \omega_{1,2} x_{2,t-\Delta t} + \omega_{2,2} x_{2,t-2\Delta t}\right] = 0 \\
&\Rightarrow x_{2,t-\Delta t} = x_{MV,t-\Delta t} = \frac{1}{\omega_{1,2}}\left(a_0 - Y^{set}\right. \\
&\quad + a_1\left[\delta_{1,1} v_{1,t+\theta_{MV}-\Delta t} + \delta_{2,1} v_{1,t+\theta_{MV}-2\Delta t} + \omega_{1,1} x_{1,t+\Delta\theta-\Delta t} + \omega_{2,1} x_{1,t+\Delta\theta-2\Delta t}\right] \\
&\quad \left. + a_2\left[\delta_{1,2} v_{2,t+\theta_{MV}-\Delta t} + \delta_{2,2} v_{2,t+\theta_{MV}-2\Delta t} + \omega_{2,2} x_{2,t-2\Delta t}\right]\right)
\end{aligned} \tag{20}$$

Eq. (20) gives an explicit solution for the FFC signal, x_{MV}, in this example. When $f(\mathbf{V})$ has terms higher than first order, numerical root solving methods may be required to find x_{MV}, as illustrated in [22].

Eq. (20) is evaluated now to determine if it can meet the standard of perfect control for x_1 load changes. For a frame of reference, MPC is also included in this study although it is a FBC method. For this example, $Y^{set} = 100$ and remains constant. Input changes are made in $x_1(t)$ and its dynamic response to these input changes, $v_1(t)$, are given in **Figure 5**. The tuning parameter for MPC, J, has values of 1, 2 and 10. The model parameters are: $a_1 = 1$, $\tau_1 = 5$ min, $\theta_1 = 5$ min; $a_2 = -1$, $\tau_2 = 10$ min, $\theta_2 = 10$ min; and the sampling time, $\Delta t = 1$ min. The results for CV and MV for both FFPC and MPC are given in **Figure 5**. FFPC gives perfect control, as anticipated, and MPC does not, as anticipated. The response of MV that gives perfect control is the heavy black line in **Figure 5**. MPC with $J = 1$ appears to match the FFPC MV profile the best in terms of shape and time of changes, but it is also the most extreme. Thus, this example illustrates the ability of FFPC to meet the requirement of theoretically perfect control. **Figure 6** gives a general multiple-input, block-oriented model FBFF block diagram similar to the one in **Figure 4**. For more information see [13].

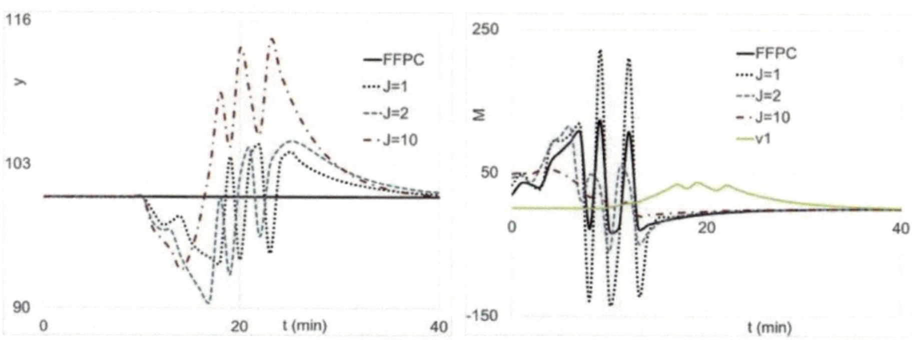

Figure 5.
CV (y) responses (left plot) and MV (M) *changes (right plot) for FFPC and MPC (J = 1, 2, and 10).* M = x_2 *for FFPC and MPC, and* M = v_1 *for the DV (i.e.,* x_1*) [13].*

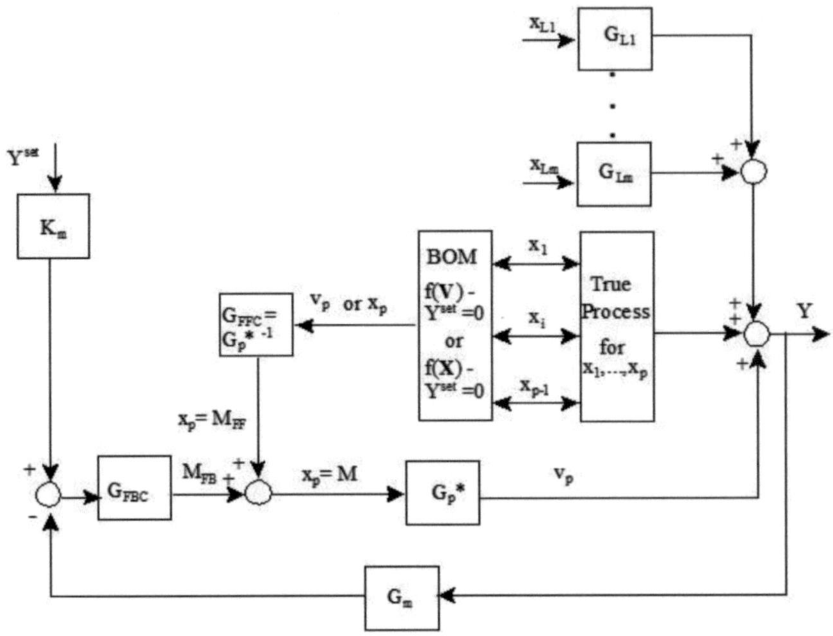

Figure 6.
A general BOM FBFF block diagram shown with m *loads and* p *FFC variables [8].*

FFPC is now evaluated on the *in silico* continuous stirred tank reactor (CSTR) in **Figure 7** and described in [13] and taken from [5] with some minor modifications. This study has two DVs – feed composition, C_{Ai} (x_1), and temperature of the coolant entering the jacket, T_{Ci} (x_2)). MV is the flowrate of the coolant entering the jacket, F_C (x_3). The output (CV) is the measured tank temperature, T_m (y). The model for each input is SOPDT, as shown in Eq. (7). The output, y, follows Eq. (8) and measurement noise was added to the true tank temperature (T) to produce T_m. Modeling this process was an application of Eqs. (9)–(17) with $\theta_1 = \theta_2 = 5$ seconds (s) and $\theta_3 = \theta_{MV} = 10$ s. Thus, $\Delta\theta = 5$ s for each input and both are PU in the FFC law. Therefore, the objective of this study is to compare announcement of input changes 5 s ahead versus no announcement.

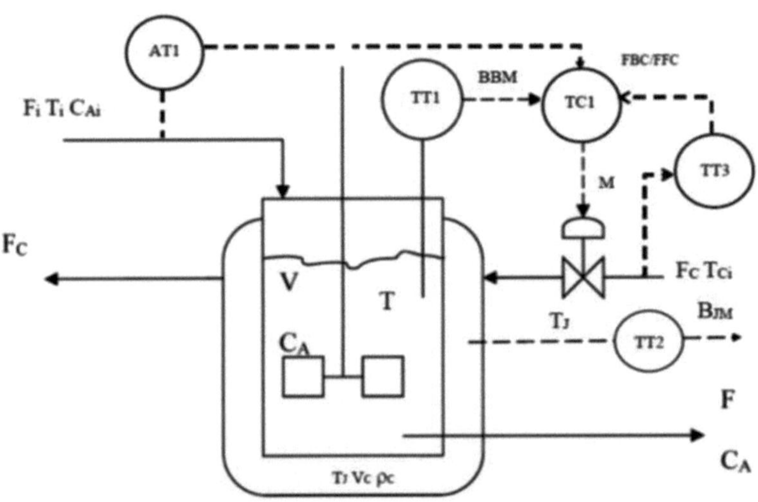

Figure 7.
Flow diagram of the CSTR in the in silico *study.*

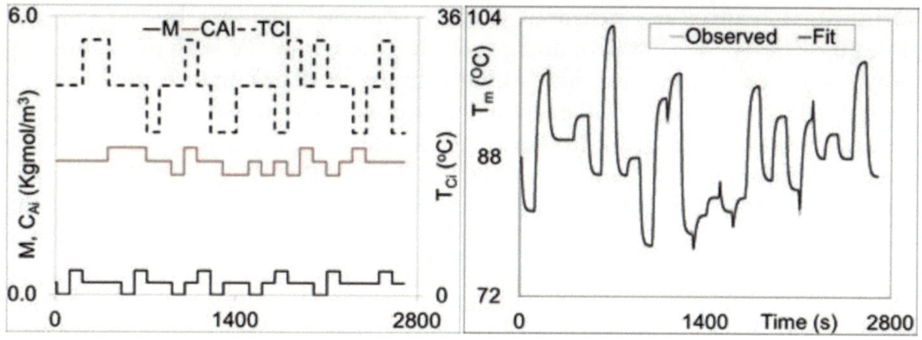

Figure 8.
Input sequences (left plot) C_{Ai} (x_1), T_{Ci} (x_2) *and M* (x_3) *and its wiener model fit (right plot).*

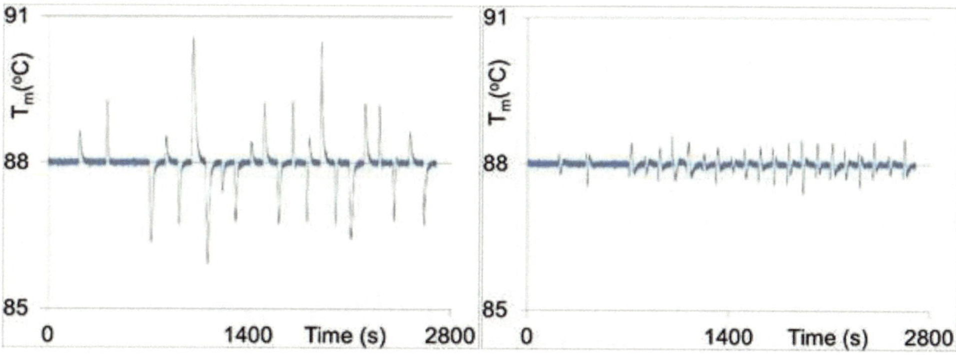

Figure 9.
The effect of T_{ci} *and* C_{Ai} *announcement on tank temperature* T_m *for FFPC (left plot is without announcement for both and right plot with announcements for both).*

A proportional-integral (PI) feedback controller was implemented in this study. Thus, FBPC was not used for FBC in any case to evaluate FFPC exclusive of FBPC. For this controller K_C = 1.40, τ_I = 11.0 and M_{FB} is the FBC signal to the control valve. $M_{FF} = x_3^e = x_{MV}^e$, is the FFPC signal. Thus, the signal to the valve, M, in **Figure 7** is $M = M_{FB} + M_{FF}$. The input sequence used for training the model is given in **Figure 8**. The excellent fit of the model to T_m for these input changes is also shown in **Figure 8**. The testing sequence (not shown) fit the response as well as the training sequence.

The results of FBC with FFPC for the two disturbances is shown in **Figure 9**. The left plot is for FBC only. The right plot is FBC with announcements for T_{Ci} and C_{Ai}. As shown, the variation of T_m around its set point decreased greatly with FFC and announcements for both disturbances. More specifically, the standard deviation about the set point temperature dropped from 0.4352°C to 0.1131°C, a 74% reduction. Thus, modeling disturbances effectively and implementing them into FFC algorithms that can take advantage of announcements of future changes for critical disturbances can have a significant impact in reducing variation around the set point of CV.

6. FBPC and the Smith predictor

As demonstrated above, FBPC is an effective FBC strategy when θ_{MV} is large. The Smith Predictor (SP) [4, 9] is a widely accepted FBC strategy when there is no deadtime in MV, and CV is measured, not at t, when MV changes, but at $t + \theta_{CV}$ (see **Figure 10**). The SP idea is to obtain a predictive model for CV with deadtime, remove the deadtime, and use this estimator in the feedback error term for CV at t. In **Figure 11** an example of a SP process is illustrated with a block diagram representing the process given in **Figure 10**. As shown in **Figure 11**, MV = $M(t)$ (M is the signal of MV) immediately affects CV = $B(t)$ (B is the signal of CV), but the sensor is at $B_1(t) = B(t\text{-}\theta_{CV})$. Thus, SP uses a prediction of CV at t with an MV that changes CV immediately. In contrast, FBPC uses a forecast prediction of CV at $t + \theta_{MV}$ when it takes a time of θ_{MV} for a change in MV to affect CV. Moreover, the SP does not use a forecast estimator and is not applicable to cases with deadtime in MV. However, this section will show that FBPC, using the forecasting estimator of CV at $t + \theta_{CV}$, gives the same result as the SP. Consequently, FBPC can be used in place of the SP. However, the opposite is not true. More specifically, the SP is not applicable when a change in MV has time delay in changing CV, whereas FBPC is applicable for this case.

For a **Figure 10** type process, the SP should compensate for the deadtime (i.e., reduce its effect) and respond quicker using an accurate estimate of B_t than using an accurate measurement of $B_{1,t}$. The block diagram for the SP [4, 9] shows feedback control using $CV_t = \hat{B}_t$ with bias correction (BC) to address measurement bias. BC is the current measurement of $B_{1,t} - \hat{B}_{1,t}$, where $\hat{B}_{1,t}$ is the estimated value of $B_{1,t}$.

Thus, in the SP block diagram,

$$e_t = Y_t^{set} - B_t - \left(B_{1,t} - \hat{B}_{1,t}\right) \tag{21}$$

Similarly, for FBPC,

$$e_t = Y_{t+\theta_{MV}}^{set} - \hat{B}_{1,t+\theta_{MV}} \tag{22}$$

The same simulated CSTR used above was used in this study to compare classical FBC, FBPC and the SP control algorithm with and without feedback correction. A

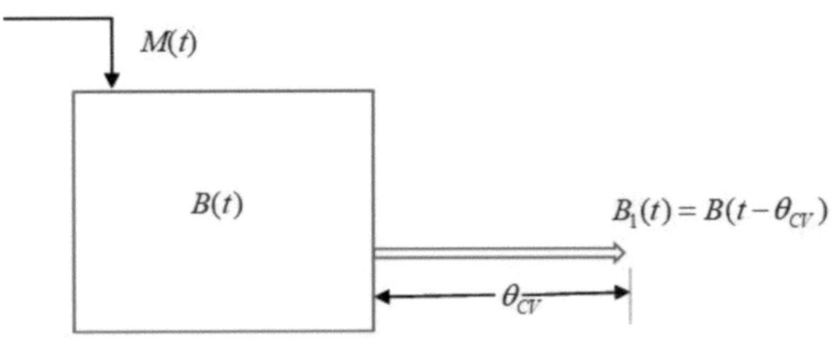

Figure 10.
An example of a SP process with M *as MV and* B_1 *as CV.*

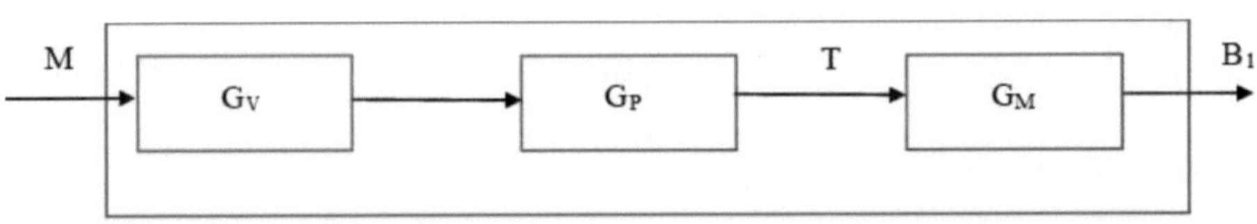

Figure 11.
A block diagram showing the blocks between M *and* B_1.

step test in M was done and obtained B_1 over time from the initial steady state to a final steady state. These values are: M_0 = 0.2569 and $M_\infty = 0.3800$ corresponding to B_{10} = 0.4000 and $B_{1,\infty} = 0.3800$. The input change was large enough to cover the change in T_m for the test data (i.e., a 4°C change in the set point temperature). A FOPDT model was fit to the data and the fitted response is given in **Figure 12**. As shown, the fit is excellent with the following estimates: $\hat{K} = 1.746,\ \hat{\tau} = 14.24\,s$ and $\hat{\theta} = 14\,s$ with $\hat{\delta} = \hat{\tau}/(\hat{\tau} + \Delta t) = 0.99303$ and Δt = 0.1. A plot of the response over time is given in **Figure 12**. The fitted forecast equation for $\hat{B}_{1,t+\theta_{MV}}$ is derived as follows:

$$\begin{gathered}\tau\frac{dB_1'(t)}{dt} + B_1'(t) = KM'(t-\theta) \qquad \Rightarrow \tau\frac{B_{1,t}' - B_{1,t-\Delta t}'}{\Delta t} + B_{1,t}' = KM_{t-\theta-\Delta t}' \\ \vdots \qquad \Rightarrow (\tau + \Delta t)B_{1,t}' = \tau B_{1,t-\Delta t}' + K\Delta t M_{t-\theta-\Delta t}' \\ \Rightarrow B_{1,t}' = \frac{\tau}{\tau+\Delta t}B_{1,t-\Delta t}' + K\frac{\Delta t}{\tau+\Delta t}M_{t-\theta-\Delta t}' \qquad \Rightarrow B_{1,t}' = \delta B_{1,t-\Delta t}' + K(1-\delta)M_{t-\theta-\Delta t}' \\ \Rightarrow \hat{B}_{1,t+140} = B_{1,t}' + 0.4 = \hat{\delta}\hat{B}_{1,t+139} + \hat{K}\big(1-\hat{\delta}\big)M_{t-0.1}\end{gathered} \tag{23}$$

The SP estimate without BC is obtained from Eq. (23) as

$$\hat{B}_{t+140} = \hat{\delta}\hat{B}_{t+139} + \hat{K}\big(1-\hat{\delta}\big)M_{t-0.1} \tag{24}$$

With BC, the SP estimate is

$$\hat{B}_t = \hat{\delta}\hat{B}_{t-0.1} + \hat{K}\big(1-\hat{\delta}\big)M_{t-0.1} - \big(B_{1,t} - \hat{B}_{1,t}\big) \tag{25}$$

For a step change in the set point temperature of 4°C, the responses for tank temperature (T_m) for all four cases are given in **Figure 13**. The proportional-integral (PI) controller is the slowest to get to the new set point of 92°C. This is no surprise since the deadtime is quite large as shown in **Figure 12**. The SP with BC gives a modest improvement over PI which is surprising. This is also reflected in the

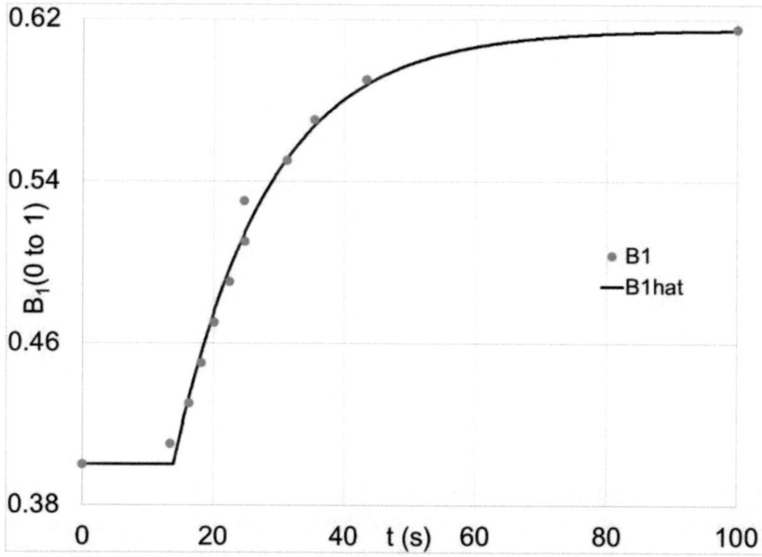

Figure 12.
The fitted process reaction curve of B_1 *the measured values used for the fitting for a step change in* M.

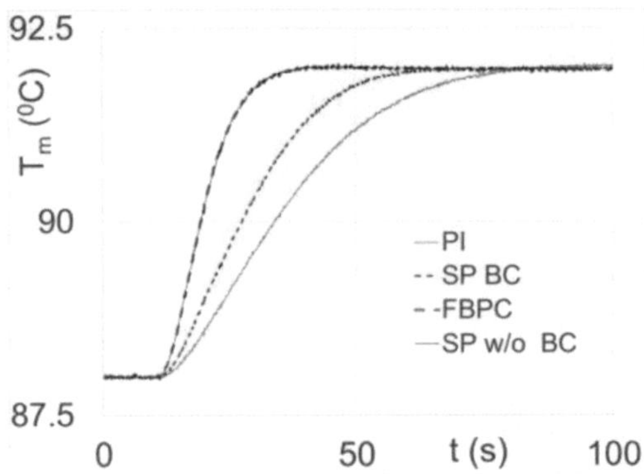

Figure 13.
Graphical SP results.

Controller	K_C	τ_I
PI	0.20	12.70
SP w BC	0.41	14.00
SP w/o BC	1.50	15.00
FBPC	1.50	15.00

Table 1.
Tuning values in the SP study.

modest increase in K_c from 0.20 to 0.41 as shown in **Table 1**. However, when the BC was removed, the SP response improved considerably as did K_c to 1.50. It gives the same response as FBPC, which also has no BC. These two cases give the same results, supporting the conclusion that FBPC and SP are equivalent for the SP application. However, when MV has deadtime with respect to CV, SP is not applicable, but FBPC is applicable. In [6] this BC method also did quite poorly with MPC being 132% worse (This BC method was not applied to FBPC in this study). Moreover, for BC, the following time series approach is recommended where the ϕ's are estimated with all other parameters for fitted model [6]:

$$\begin{aligned}\hat{y}_{t+\theta} &= \hat{\eta}_{t+\theta} + \hat{\phi}_1(y_t - \hat{\eta}_t) + \hat{\phi}_2(y_{t-\Delta t} - \hat{\eta}_{t-\Delta t}) + \cdots \\ &= \hat{\eta}_{t+k_1\Delta t} + \hat{\phi}_1 e_t + \hat{\phi}_2 e_{t-\Delta t} + \cdots\end{aligned} \tag{26}$$

7. Conclusions

This chapter has focused on the use of discrete-time dynamic forecast modeling to enhance FBC and all types of FFC. Discrete-time modeling has the advantage of obtaining solutions to PU systems without having to make assumptions to make the system an approximation of a physically realizable system. Models do not have to be *cause-and-effect* for forecasting but need to be as FFC models. Cause-and-effect models result from statistical design of experiments because input changes are orthogonal (i.e., uncorrelated) and for theoretical structured models because they will be nonlinear in one or more physically based parameters, have physical constraints that must be met, and physically interpretable unknown model parameters [23].

When MV has deadtime with-respects-to CV (e.g., θ_{MV}), a change in MV will not begin to change CV until a time distance of θ_{MV} in the future. Three FBC approaches where evaluated in this scenario: classical FBC, FBPC and MPC, using the current measured value of CV, forecast estimate of CV, θ_{MV} in the future, and forecast estimate of CV, $\theta_{MV} + J\Delta t$ in the future, respectively. In the simulation study, FBC responded quicker than MPC when the process lag was large and MPC responded quicker when the process lag was small. FBPC responded much faster than both under small and large lag. FBPC control has a prediction horizon of θ_{MV} but for MPC it is $J\Delta t$ longer. Since the optimal value of J increases as the lag increases, MPC can be significantly more sluggish than FBPC when J is large. A definite advantage of MPC is that J is its only tuning parameter.

A discrete-time FFC approach (FFPC) was presented in this chapter that can be effective when θ_{MV} is large and the multiple-input FFPC model is PU for any reason (i.e., the order of the differential equation or negative deadtime). FFPC

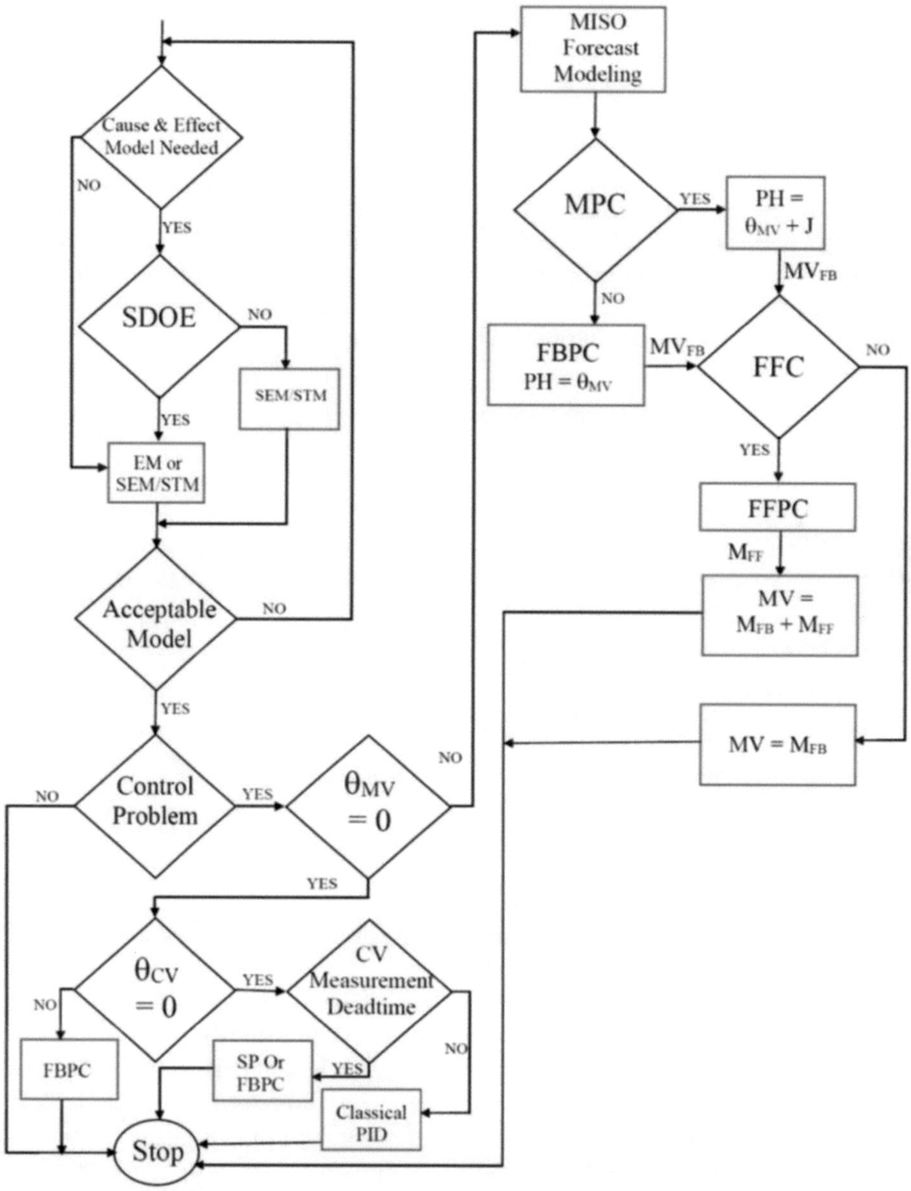

Figure 14.
Flowchart illustrating the complete process of the proposed framework of this chapter.

was shown to satisfy perfect theoretical control in a simulated data study. A critical strength of the approach presented in this work is that the FFC variables enter one mathematical function that simultaneously solves for one FFC control signal. This contrasts with classical FFC that has a FFC algorithm for each input and combines their values to determine the value of the FFC control signal for MV. The classical approach cannot treat complex interactive and nonlinear behavior of the disturbances in determining the optimal value of the FFC signal for MV. Block diagrams of this novel FFC approach were shown for the Wiener Network and a general block-oriented modeling approach. When FFC inputs have a PU impact, knowing how their values will change over the control horizon (i.e., announcements), can significantly improve FFC as demonstrated in the CSTR simulation study.

The SP is a model-based feedback control algorithm that can be quite effective when there is no deadtime between a change in MV and its impact on CV, and the measured value of CV has deadtime. For this situation, FBPC, that uses a forecast value for CV based on a model developed from the measured value of CV with deadtime, gives the equivalent result of the SP. However, the SP is limited to this case, but FBPC is not. More specifically, FBPC is applicable when there is deadtime for changes in MV and its effect on CV but the SP is not. Finally, one should exercise care when using the bias correction (BC) method in the block diagram for the SP. It can lead to a significantly suboptimal SP as shown in this work. A better alternative is to use one that is obtained from modeling the serially correlated structure in the process as given in this chapter. A flowchart illustrating the complete process of the proposed framework of this chapter is shown in **Figure 14**.

Acknowledgements

I thank Kendra Kreienbrink for helping with the references and for proofreading this document.

Nomenclature

A	cross-sectional area of the tank
a_n	the nth constant
B	the signal of CV in the SP algorithm
BC	bias correction in the SP algorithm
B_1	the signal of CV with deadtime θ_{CV}
b_m	the mth constant
C_{Ai}	inlet concentration to the CSTR (x_1)
e_{ffc}	the feedforward control law criterion for perfect control
e_t	the feedback error = $Y^{set} - y_t$,
F_C	inlet flow rate of the coolant to the jacket of the CSTR
$f(V^e)$	the true function in the FFC law that satisfy $e_{ffc} = 0$
$f(\boldsymbol{V})$	is an unrestricted mathematical function that maps each v_i' to the units of the output variable in standard form. Thus, it is $f(\boldsymbol{V})$ that transforms the linear dynamic inputs into the nonlinear dynamic response for the output y.
G_f	feedforward transfer function
G_{DV}	DV transfer function
G_i	transfer function with output v_i

G_M transfer function for CV in signal
G_{MV} MV transfer function
G_P process transfer function
G_V transfer function for MV
h the tank level
J the only controller tuning parameter for MPC.
K process gain
K_C controller gain for FBC
m m^{th} order derivative
M the input signal to MV transfer function
M_{FB} FBC input signal to MV transfer function
M_{FF} FFC input signal to MV transfer function
n n^{th} order derivative
q_1 inlet flow rate of Stream 1
Q flowrate in the Laplace domain entering the system
q flow rate of the outlet stream
R_v valve resistance
t current time or just time
T the physical value of CV in SP case
T_{Ci} inlet temperature of the coolant to the jacket of the CSTR
T_m measured tank temperature
v_i dynamic output for i^{th} input, x_i, in the same units as x_i
$V(t)$ is a vector of the $v_i''s$
x input
$x_{t-\Delta t}$ the value of x at time t -Δt
x_i' deviation of x_i at time t from x_i at time t = 0
x_p^e x_{MV}^e equals the value of MV in the FFC law that satisfy $e_{ffc} = 0$
y the response = output
$\hat{y}(t)$ the estimate of $y(t)$

Greek Letters

θ time delay or deadtime
$\eta(t)$ the true value of the output at time t
$\hat{\eta}(t)$ the estimate of $\eta(t)$
$\omega_{i,j}$ dynamic coefficient for the ith input at $t - j\Delta t$
$\delta_{i,j}$ dynamic coefficient for output v_i at $t - j\Delta t$
θ_{CV} dead time of the controlled variable
θ_{DV} dead time of the disturbance variable
θ_{MV} dead time of the manipulated variable
τ_I reset time for FBC
τ process time constant
Δt the sample rate

Acronyms and abbreviations

BGC blood glucose concentration
CSTR Continuous-Stirred-Tank-Reactor
CV control variable
DT deadtime
DV disturbance variable
EM empirical modeling

EMM	empirical modeling methods
FBC	feedback control
FBFFC	feedback/feedforward control
FBPC	feedback predictive control
FFC	feedforward control
FFPC	feedforward predictive control
FOPDT	first-order-plus-deadtime
MISO	multiple-input, single-output
MPC	model predictive control
MV	manipulated variable
PID	proportional, integral, derivative
PH	prediction horizon
PU	physically unrealizable
SEM	semi-empirical model
SDOE	statistical design of experiments
STM	semi-theoretical model
SP	Smith Predictor
SOPDT	second-order-plus-deadtime
TDPU	time delay physical unrealizability

Author details

Derrick K. Rollins
Department of Chemical, Biological Engineering and Statistics, Iowa State University, Ames, Iowa, USA

*Address all correspondence to: drollins@iastate.edu

References

[1] Mellodoge P. A Practical Approach to Dynamical Systems for Engineers. Ltd: Elsevier; 2016

[2] Tan L, Jiang L. Digital Signal Processing. 3rd ed. Inc: Elsevier; 2019

[3] Aldrich J. Correlations genuine and spurious in Pearson and Yule. Statistical Science. 1995;**10**(4):364-376

[4] Seborg, D, Edgar T, Mellichamp D. Process Dynamics and Control, 2nd edition, Wiley 2004.

[5] Smith C. Corripio A. Principles and Practice of Automatic Process Control: John Wiley & Sons; 1985

[6] Rollins D, Mei Y. A new feedback predictive control approach for processes with time delay in the manipulated variable. Chemical Engineering Research and Design. 2018; **136**:806-815

[7] García C, Prett D, Morari M. Model predictive control: theory and practice – a survey. Automatica. 1989;**25**(3): 335-348

[8] Quin S, Badgwell T. A survey of industrial model predictive control technology. Control Engineering Practice. 2003;**11**(7):733-764

[9] Smith O. Closer control of loops with dead time. Chemical Engineering Progress. 1957;**53**(5):217-219

[10] Huyett L, Dassua M, Zisser H, Doyle F. Design and evaluation of a robust PID controller for a fully implantable artificial pancreas. Industrial & Engineering Chemistry Research. 2015;**54**:10311-10321

[11] Rae-Puckett W. Dynamic modelling of diabetes mellitus [Dissertation]. Wisconsin-Madison: University of Wisconsin-Madison; 1992

[12] Noguchi C, Hashimoto S, Furutani E. In silico blood glucose control for type 1 diabetes with meal announcement using carbohydrate intake and glycemic index. Advanced Biomedical Engineering. 2016;**5**:124-131

[13] Clarke W, Anderson S, Breton M, Patek S, Kashmer L, Kovatchev B. Closed-loop artificial pancreas using subcutaneous glucose sensing and insulin delivery and a model predictive control algorithm: the Virginia experience. Journal of Diabetes Science and Technology. 2009;**3**(5):1031-1038

[14] Pearson R, Pottmann M. Gray-box identification of block-oriented nonlinear models. Journal of Process Control. 2000;**10**(4):301-315

[15] Rollins D, Bhandari N, Kleinedler J, Kotz K, Strohbehn A, Boland L, et al. Free-living inferential modeling of blood glucose level using only noninvasive inputs. Journal of Process Control. 2010;**20**:95-107

[16] Kotz K, Cinar A, Mei Y, Roggendorf A, Littlejohn E, Quinn L, et al. Multiple-input subject-specific modeling of plasma glucose concentration for feedforward control. Ind. Eng. Chem. Res. 2014;**53**(47):18216-18225

[17] Rollins D, Mei Y, Kotz K, Littlejohn E, Quinn L, Roggendorf A, et al. An extended static and dynamic feedback/feedforward control algorithm for insulin delivery in the control of blood glucose level. Ind. Eng. Chem. Res. 2015;**54**(26):6734-6748

[18] Cameron F, Niemeyer G, Bequette B. Extended multiple model prediction with application to blood glucose regulation. Journal of Process Control. 2012;**22**:1422-1432

[19] Bequette B. Algorithms for a closed-loop artificial pancreas: the case for

model predictive control. Journal of Diabetes Science and Technology. 2013; 7(6):1632-1643

[20] Wang Q, Xie J, Molenaar P, Ulbrecht J. Model predictive control for type 1 diabetes based on personalized linear time-varying subject model consisting of both insulin and meal inputs: an in silico evaluation. J Diabetes Sci Technol, 2015; Jul, 9(4), 941-942.

[21] Marchetti G, Barolo M. and etc. A feedforward-feedback glucose control strategy for Type 1 Diabetes Mellitus. Journal of Process Control. 2008;**18**(2): 149-162

[22] Rollins D, Mei Y, Loveland S, Bhardari N. Block-oriented feedforward control with demonstration to nonlinear parametrized wiener modeling. Chemical Engineering Research and Design. 2016;**109**:397-404

[23] Rollins D, Roggendorf A, Khor Y, Mei Y, Lee P, Loveland S. Dynamic modeling with correlated inputs: theory, method and experimental demonstration. Ind. Eng. Chem. Res. 2015;**54**(7):2136-2144

Chapter 5

Optimization of Model Predictive Control Weights for Control of Permanent Magnet Synchronous Motor by Using the Multi Objective Bees Algorithm

Murat Sahin

Abstract

In this study, the model predictive control (MPC) method was used within the scope of the control of the permanent magnet synchronous motor (PMSM). The strongest aspect of the MPC, the ability to control multiple components with a single function, is also one of the most difficult parts of its design. The fact that each component of the function has different effects requires assigning different weight coefficients to these components. In this study, the Bees Algorithm (BA) is used to determine the weights. Using the multi-objective function in BA, it has been tried to determine the weights that reduce the current values together with the speed error. Three different PI controllers have been designed to compare the MPC method. The coefficients of one of these are tuned with BA. Good Gain Method and Tyreus-Luyben Method were used in the other two. As a result of experimental studies, it has been observed that MPC can control PMSM more smoothly and accurately than PI controllers, with weights optimized with BA. With MPC, PMSM has been controlled with 15% settling time than other controllers and also with no overshoot.

Keywords: Model predictive control, permanent magnet synchronous motor, the Bees Algorithm, the Good Gain method, Tyreus-Luyben method

1. Introduction

Permanent Magnet Synchronous Motors (PMSM) have been used for many years due to their features such as high torque, high efficiency and fast dynamic structure. Within the scope of controlling PMSM; robust control [1], field-oriented control and direct torque control [2], fuzzy-based controllers [3, 4], sliding mode controller [5], model predictive controller (MPC) [6], and so on different control methods have been used. Especially in high speed PMSMs, driver dynamics must be controlled successfully for effective control [7]. When the applications in the literature are examined, it is seen that MPC gives successful results in this scope. Model Predictive Control (MPC), based on the optimal control theory, achieves successful results, especially in power electronics applications. MPC uses the system model equations together with the current state measurements to estimate the control movement. With this predictive ability, it can achieve more successful results against traditional controllers [8].

Three phase inverter circuits are one of the main methods used to drive PMSM. In these circuits, there can be a limited number of switching combinations for 6 switches. This is called the Finite Control Set (FCS). FCS-MPC can be implemented even in low cost devices used today due to the optimization process performed with a limited number of iterations [9]. One of the major advantages of the FCS-MPC over other control methods is that different goals, variables, and constraints can be included in a single cost function and controlled simultaneously [10]. Adding different categories of variables to the cost function brings great flexibility to MPC. However, the effects of these variables on the system may be different. Therefore, it is necessary to give weight coefficients for each of the variables.

Selection of weights for cost function of MPC optimization is one of the main challenge for researchers. When the MPC studies in the literature are examined, it is seen that a significant part of them are about to calculate these weights. It is seen that different methods are used in different applications. The highlights of these; empirical methods, fuzzy-based methods, evaluation algorithms, heuristic methods etc. In a sample study, a fuzzy-based calculation method was used in the PMSM current control application. Id and Iq currents were used in the cost function [11]. In another study, torque control of PMSM with MPC was performed. Torque and flux variables were used in the cost function. While weight was not used for the torque variable, a weight depending on the torque has been determined for the flux variable [12]. A different cost function that can be selected for the speed control of the PMSM may include the controller output and speed error. In order not to cause sudden effects on the system, the difference of the two control signals produced consecutively is added to the cost function [13]. A similar strategy has been used in DC motor control. With the Quadratic problem approach, along with the speed error and the difference between the two consecutive outputs are included in the cost function [14].

FCS-MPC is also used in different fields other than electric motors and similarly the weights need to be calculated. Simulated Annealing Particle Swarm Optimization with Model Predictive Control was used to control of the electric vehicles [15]. Another application that uses genetic algorithm is shunt active power filter. Link voltage, active power and reactive power are used in cost function of MPC within the scope of proper switching [16]. In a different MPC application, the multi objective genetic algorithm was used to calculate the weights [17].

In this study, the control of the PMSM used as the driving element of an actuator to be used in the aerospace area was performed with FCS-MPC. Battery life is of great importance in the aerospace area. For this reason, when creating the MPC cost function, the power variable was added along with the speed and current variables. In the PMSM control problem, apart from minimizing the speed error, minimum current consumption is also aimed. In the scope of calculating the weights, the multi objective BA is used. When the studies in the literature are analyzed, it is seen that BA gives successful results against multi objective function problems [18–20]. In order to compare the developed control system, 3 different PI controllers were designed. The second part of the study includes PMSM equations and MPC studies. The third part includes calculating weights with multi-objective BA studies. In the fourth chapter, there are studies on PI controller design. The fifth section includes experimental studies and comparisons.

2. PMSM equations and FCS-MPC design

In MPC design, first of all, it is necessary to prepare the mathematical model of the system. One of the most popular methods of controlling of PMSM with MPC is to use machine equations in the rotor reference frame [6]. MPC is designed within the scope of speed control of PMSM. Therefore, speed and current equations were needed in the MPC cost function.

$$\frac{di_d}{dt} = -\frac{R}{L_d} i_d + \frac{L_q p}{L_d} w_r i_q + \frac{1}{L_d} v_d \tag{1}$$

$$\frac{di_q}{dt} = -\frac{R}{L_q} i_q - \frac{L_d p}{L_q} w_r i_d - \frac{\psi_m p}{L_q} w_r + \frac{1}{L_q} v_q \tag{2}$$

$$T_e = \frac{3}{2} p \left[\psi_m i_q + (L_d - L_q) i_d i_q\right] \tag{3}$$

$$\frac{dw_r}{dt} = \frac{1}{J}(T_e - T_l) - \frac{B}{J} w_r \tag{4}$$

In the currents and torque equations; ψ_m, i_q, i_d, v_q, v_d, R, L_q, L_d, w_r, and p and are the rotor magnetic flux linkage, stator currents in q and d axis, stator voltages in q and d axis, stator resistance, stator inductances in q and d axis, rotor angular speed, and pole pairs respectively. In the equation of speed; J is the inertia, T_l is the load torque, B is the viscous friction coefficient, and T_e is the electrical torque produced by the motor [11].

In the equations, resistance, inductance, pole pairs, inertia, and magnetic flux values are known. (Motor resistance and inductance can vary depending on the motor temperature [21]. This situation has been neglected in this study. It should be consideration in applications where the motor will operate under load for a long time.) In order to find Vd and Vq values, the inverter circuit must be analyzed. A typical 3 Phase Inverter circuit used in the study is given in **Figure 1**. The circuit has two driver components for each phase of the PMSM and switches for switching these components. (In this study, mosfet is used as the driver component, it is shown in green color in the figure.) In order that the components on the same phase are not switched at the same time, there is a note gate between them. Therefore, 3 switches (Switches S_a, S_b, and S_c shown in blue color in the figure.) are sufficient for switching operations. These 3 switches can be switched in 8 different combinations, each with a 0 or 1 [22]. Conversions of V_{dc} (Bus voltage) to V_d and V_q according to the switching states are found by Park-Clarke methods [23].

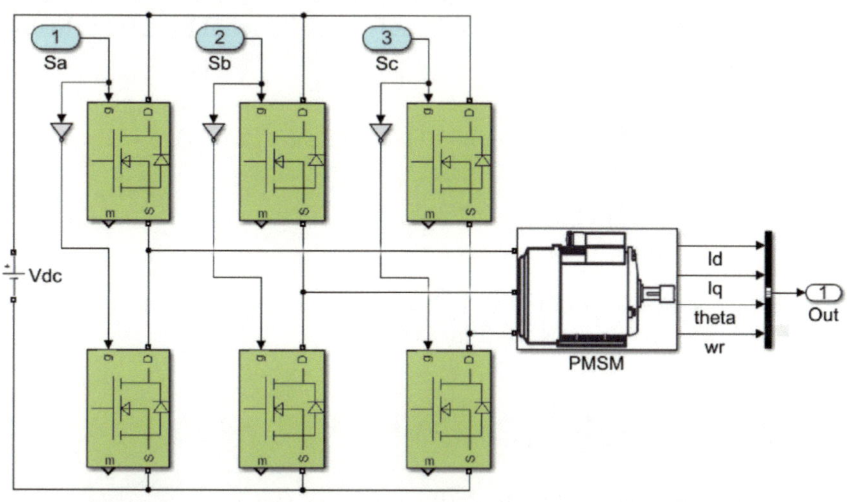

Figure 1.
PMSM drive with 3 phase inverter.

$$\begin{bmatrix} v_d \\ v_q \end{bmatrix} = \frac{2}{3} V_{dc} \begin{bmatrix} cos\theta & sin\theta \\ -sin\theta & cos\theta \end{bmatrix} \begin{bmatrix} 1 & -1/2 & -1/2 \\ 0 & \sqrt{3}/2 & -\sqrt{3}/2 \end{bmatrix} \begin{bmatrix} S_a \\ S_b \\ S_c \end{bmatrix} \quad (5)$$

θ is the angle of rotation of the rotor in radians (In this study, it is measured by the encoder integrated into the PMSM. Also, w_r is obtained by the derivative of θ.). Other unknowns of the equations are the I_d and I_q currents. I_a, I_b, and I_c currents are measured with the current sensor. Again, Park-Clarke method is used for transition from abc phases to dq axis (In the Eq. (5), instead of switches, this time the currents are placed.).

Various discretization methods can be used to obtain a discrete-time model for calculating predictions. One of the simplest methods is the Forward Euler method, which is based on derivatives. In this method, the prediction expression is obtained by leaving the expression x(k + 1) alone. T_s is the sampling time [23].

$$\frac{dx}{dt} \approx \frac{x(k+1) - x(k)}{T_s} \quad (6)$$

When the Forward Euler approach is applied to stator currents (1) and (2) the following MPC prediction equations are obtained [11].

$$i_d(k+1) = \left(1 - \frac{RT_s}{L_d}\right) i_d(k) + \frac{L_q}{L_d} T_s p w_r i_q(k) + \frac{T_s}{L_d} v_d \quad (7)$$

$$i_q(k+1) = \left(1 - \frac{RT_s}{L_q}\right) i_q(k) - \frac{L_d}{L_q} T_s p w_r i_d(k) - \psi_m p w_r \frac{T_s}{L_q} + \frac{T_s}{L_q} v_q \quad (8)$$

In the equations, expressions with (k) show the values measured from PMSM at the previous sampling time, while the expressions with (k + 1) show the predicted values.

For each sampling time, for the 8 different Vd and Vq values given above, currents estimates will be made. Under normal conditions, what is expected from the microprocessor is to make these calculations in the period determined for the application and to generate the necessary control signal. In real application, the microprocessor also has many different tasks. Therefore, there may be delays in estimates. In this case, a dynamic system cannot be controlled successfully. Against these possible delays, it has been suggested to predict two next steps. In this study, the prediction equations for the next two steps were updated by taking this suggestion into consideration [24].

$$i_d(k+2) = \left(1 - \frac{RT_s}{L_d}\right) i_d(k+1) + \frac{L_q}{L_d} T_s p w_r i_q(k+1) + \frac{T_s}{L_d} v_d \quad (9)$$

$$i_q(k+2) = \left(1 - \frac{RT_s}{L_q}\right) i_q(k+1) - \frac{L_d}{L_q} T_s p w_r i_d(k+1) - \psi_m p w_r \frac{T_s}{L_q} + \frac{T_s}{L_q} v_q \quad (10)$$

For velocity prediction, Eq. (4) is discretized by Forward Euler method. In the equation, $i_q(k+2)$ expression is used in T_e [6].

$$w_r(k+1) = \left(1 - \frac{B}{J}\right) w_r(k) T_s + \frac{1}{J}(T_e - T_l) T_s \quad (11)$$

For velocity control in PMSM, Eqs. (9), (10) and (11) may be sufficient in the cost function. But for a more effective cost function, it has been proposed to include the switching losses in the inverter circuit. In the sample study, the power variable was also added in this context, and the choices were made to use low power at each step of the control system [25]. In this study, the power effect (P_f) was added to the cost function. However, this effect is added as shown in (12) in a simple form so that the processing load does not increase.

$$P_f = \sqrt{(v_d * i_d(k+2))^2 + (v_q * i_q(k+2))^2} \tag{12}$$

One of the strong features of MPC is that the constraints required for the system can be defined in cost function. The combination that gives the best result among the switching alternatives may also cause high current from the PMSM. Therefore, the current constraint given in (13) has been added to the cost function [22]. When the current values are higher than a limit value, a large value is assigned to the relevant switching option to be excluded from the options. (In the part indicated by ∞ in the equation, 1e10 is used in the application.)

$$\hat{f}(i_d(k+2), i_q(k+2)) = \begin{cases} \infty \; if \; |i_d| > i_{max} \; or \; |i_q| > i_{max} \\ 0 \; if \; |i_d| \leq i_{max} \; and \; |i_q| \leq i_{max} \end{cases} \tag{13}$$

The final version of the cost function is shown below.

$$g = \boldsymbol{w_1} * (w_{ref} - w_r(k+1))^2 + \boldsymbol{w_2} * (i_d(k+2))^2 + \boldsymbol{w_3} * (i_q(k+2))^2 + \boldsymbol{w_4} * P_f{}^2 + \hat{f}(i_d(k+2), i_q(k+2)) \tag{14}$$

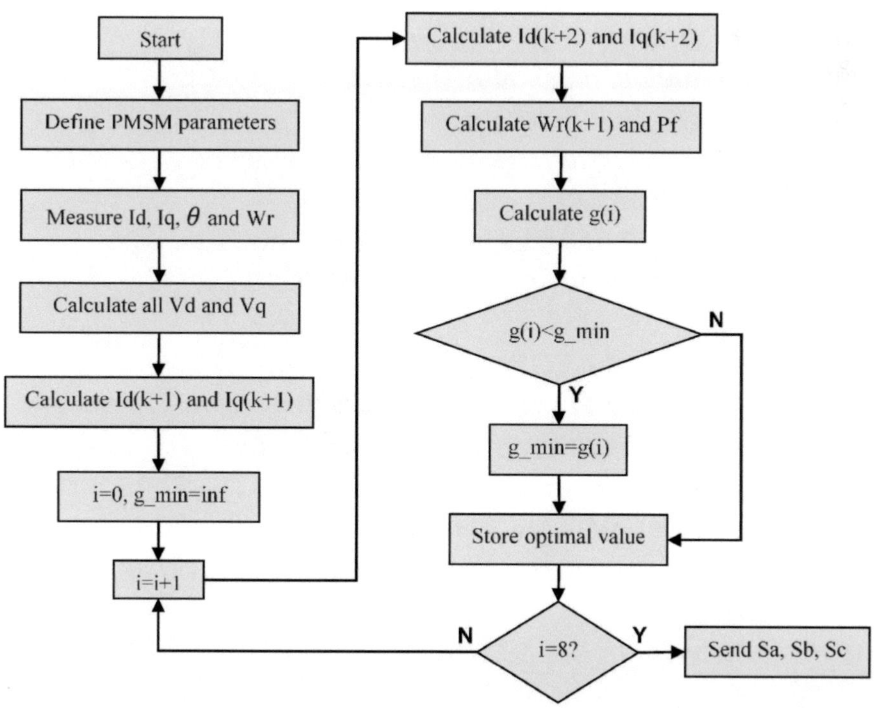

Figure 2.
Flowchart of MPC.

For each sampling time, for 8 different switching combinations, current and velocity prediction will be performed and the cost function defined in (14) will be calculated. After 8 iterations, the switching configuration that gives the minimum "g" value will be determined and sent to the inverter circuit. The flow diagram of the designed MPC is given in **Figure 2**.

3. Calculating weights with multi objective Bees Algorithm (MOBA)

BA is a population-based search algorithm. The algorithm mimics the nectar source search behavior of honey bees. Basically, it does some kind of neighbor region search along with random search and can be used for both integrated and functional optimization [26]. Detailed explanations about the algorithm were provided in Ref.s [27–29]. The pseudo code of the algorithm is given in **Figure 3**.

When designing control systems in general, the integral of the square of the error (ISE) is used as the objective function. The equation of ISE is given in (15). The equation shows the reference velocity value with r(t), the output velocity value with y(t) and the error value with e(t).

$$ISE = \int_0^{\infty} (r(t) - y(t))^2 dt = \int_0^{\infty} e(t)^2 dt \tag{15}$$

When the error is only aimed to be minimized, it may cause high currents to be drawn from the PMSM and thus excessive energy consumption. As a solution to this issue, a multi objective (MO) optimization algorithm is suggested. The goal of MO optimization is to try to optimize all defined objective functions simultaneously. All objectives can be minimized or maximized at the same time, or some can be minimized and some maximized. General definition in the literature is given below [30].

```
Generate random population and evaluate fitness.
Sort population
While (stopping criterion not met)
    For i=1:EliteSite
        Recruit bees for elite site and evaluate fitness
        Select the best bee from each site
    End
    For i=EliteSite+1:BestSite
        Recruit bees for best site and evaluate fitness
        Select the best bee from each site
    End
    For i= BestSite+1:ScoutBee
        Generate random population and evaluate fitness.
    End
    Sort population
End While
```

Figure 3.
Pseudo code of BA.

$$Min\ (or\ Max)\left\{ f_1(x) = y_1, f_2(x) = y_2, \dots, f_j(x) = y_j \right\} \tag{16}$$

Although there are different MO methods, the prominent method is the weighted sum method. Here, each goal has a weight coefficient. This method is also called scalarization method. In this method, basically, multiple solutions are combined into a single solution using weights [31].

$$Min(or\ Max)\sum_{j=1}^{N} w_j f_j(x) \tag{17}$$

It is necessary to limit the current for low power consumption [32]. Based on this situation, the integral of the square of the i_{bus} is included in the multi objective function (MOF) to find weights that will also minimize the i_{bus} current.

$$MOF = \int_0^{\infty} \left(e(t)^2 + i_{bus}(t)^2 \right) dt \tag{18}$$

The optimization process consists of two main parts. First part is Matlab M file with MOBA, second part is Simulink file with MPC, 3 Phase Inverter and PMSM models. The algorithm starts with the definition of BA parameters given in Appendix-B. Using these parameters, a random first population is created. Simulink model is run and MOF value is obtained for each member of the population. After this process is completed, the first population is sorted from small to large, according to the MOF value. Then local search section starts. In the elite and non-elite local areas, new values are generated by neighborhood search and simulations are made with these values in the Simulink model. In the global search, new sites are discovered randomly. Finally, the new population is re-sorted. These operations are repeated for all iterations and the best values are recorded when completed.

The system model used in the simulation is given in **Figure 4** and the flow chart of the optimization algorithm is given in **Figure 5**. The simulation model has been prepared in discrete-time to be close to the real application. The sampling frequency is 50KHz. (Information about the model is given in Appendix-A.) The w1, w2, w3, and w4 weights produced by the optimization algorithm are transmitted to the Simulink model and the simulation is performed. At the end of the simulation, the outputs are taken with the "Error" and "Current" blocks and sent to the MOF in the optimization algorithm.

One of the critical parameters in the optimization algorithm is determining the value ranges of weights. BA focuses on the areas with the best values with the first iterations. Therefore, even if large ranges are specified for variables, it quickly shrinks the solution set to include the parts with the best. For this reason, when

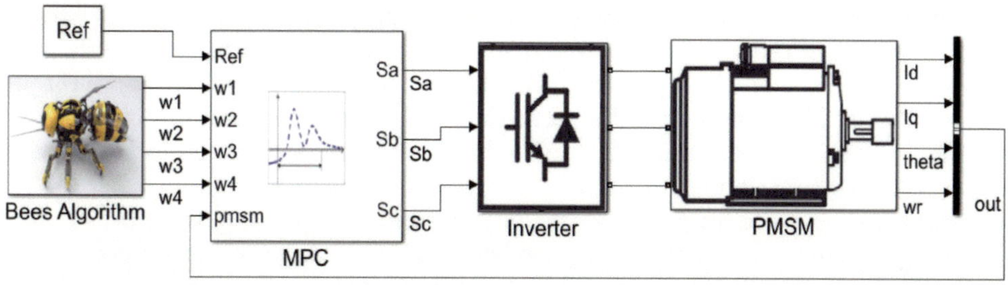

Figure 4.
Model of PMSM control with MPC.

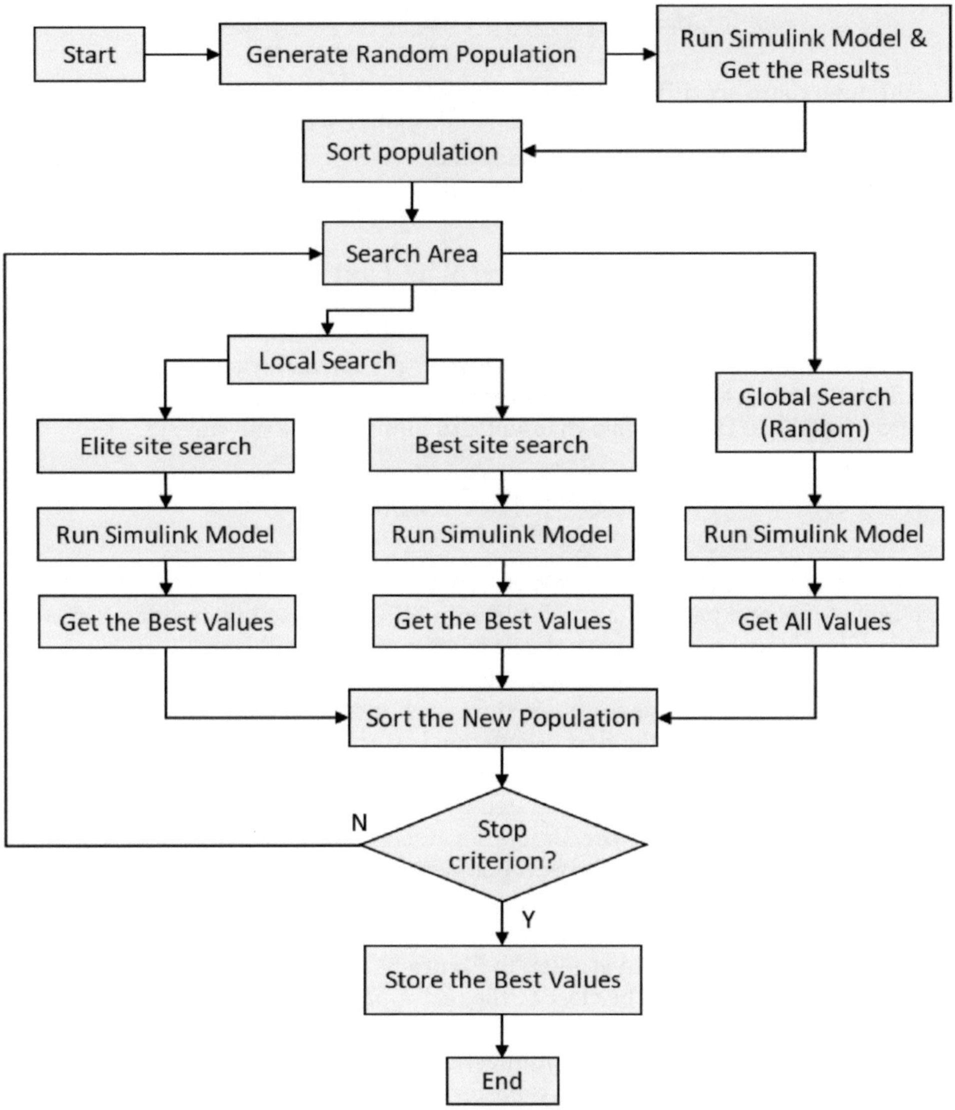

Figure 5.
Flowchart of optimization algorithm.

determining the value ranges, the range 0–1000 was initially chosen to have a wide solution range. After the optimization study these values were calculated; w1 = 251.5, w2 = 6.9, w3 = 5.1, and w4 = 1.05. Consideration of these values, 300 for w1 and 10 for the others were selected and a second optimization study was carried out. The results of the both studies are given in **Figure 6**. As can be seen, in the first optimization where wide ranges are defined, there are higher error values in the first iterations. But along with other iterations, cost functions are minimized quickly. In the second optimization, since the limit values are chosen in a narrower ranges, the cost function change is also in a narrow area. (Sufficient number of searches must be made for escaping the local minimums. The important parameters in this regard are the number of foraging bees and the number of iterations. As can be seen from the figures, the minimization process has been fixed in the last iterations. This indicates that the current algorithm parameters are sufficient. If the decline continues in the last iterations, it is necessary to update the parameters.)

It	w1	w2	w3	w4	Cost
1	938,6396	734,2122	22,7597	106,3132	20,8203
2	917,7667	665,6631	24,3344	82,1057	17,4343
3	834,7710	518,1121	20,0664	64,2322	15,5069
4	888,4555	325,1831	19,4840	45,0660	11,3746
5	714,9332	161,1120	21,3138	29,3554	10,1064
6	544,4434	104,6522	14,8776	21,8344	9,9240
7	359,9001	77,4331	15,5587	12,3356	9,3730
8	228,0098	9,9960	5,3908	2,0361	7,6221
9	233,2833	4,9776	5,5422	1,4421	7,6142
10	251,5511	6,9205	5,1322	1,0520	7,5858
11	251,5511	6,9205	5,1322	1,0520	7,5858
12	251,5511	6,9205	5,1322	1,0520	7,5858
13	251,5511	6,9205	5,1322	1,0520	7,5858
14	251,5511	6,9205	5,1322	1,0520	7,5858
15	251,5511	6,9205	5,1322	1,0520	7,5858
16	251,5511	6,9205	5,1322	1,0520	7,5858
17	251,5511	6,9205	5,1322	1,0520	7,5858
18	251,5511	6,9205	5,1322	1,0520	7,5858
19	251,5511	6,9205	5,1322	1,0520	7,5858
20	251,5511	6,9205	5,1322	1,0520	7,5858

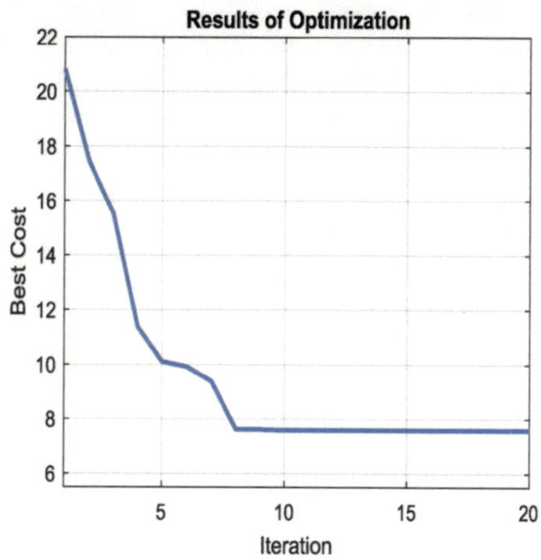

Figure 6.
Results of optimization (MPC).

4. PI controller design

One of the advantages of controlling PMSM with MPC that it does not need an external section for commutation. When controlling PMSM with PID control, it is necessary to prepare a commutation section as well. One of the methods used in this context is PWM (Pulse width modulation). The PWM signal provides the signal in a certain order with the duty cycle changes so that the DC signal becomes an AC signal. (If this signal is passed through a low-pass filter, a pure sine wave is obtained.) In SPWM (Sinusoidal Pulse Width Modulation), two signals are compared. The reference signal is sinusoidal and the carrier signal is triangular. Pulses are produced by comparing two signals, and the width of each pulse varies in proportion to the amplitude of the sine signal. The frequency of the reference signal determines the inverter output frequency and controls the reference peak amplitude, the modulation index of the output voltage, and the RMS value [33]. SPWM model used in the simulation is given in **Figure 7**.

Simulink model prepared for PI controller design is given in **Figure 8**. Id and Iq currents are used in the model as in MPC simulation. As can be seen from Eq. (3),

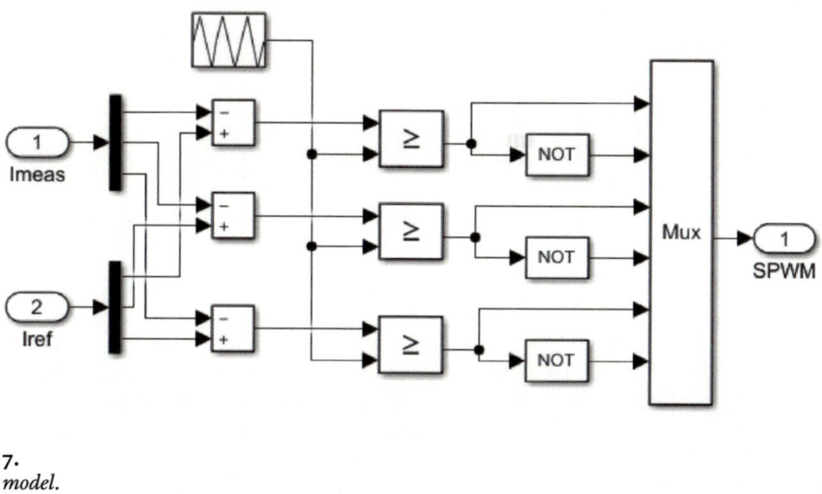

Figure 7.
SPWM model.

the PMSM speed depends on the electrical torque generated and hence the Iq and Id currents. If the inductances Lq and Ld are the same or very close, the electrical torque depends only on the Iq current. As can be seen from the motor parameters given in the Appendix-A, the Lq and Ld values are equal. Therefore, Iq current is used to perform speed control. The output of the speed controller drives the current Iq. SPWM is created for the calculated current value and PMSM is controlled through the Inverter. The sampling time and solver type of the model are the same as the MPC model.

The coefficients of the PI controller are first tuned with BA. The same structure in **Figure 5** is prepared for the PI model. Similarly, the error value is used in the objective function together with the currents. For Kp and Ki, values from 0 to 10000 were set as the limit for a wide range. BA parameters are the same as in the MPC study. **Figure 9** shows the optimization results. Costs decrease with the first iteration. The minimum value was reached with the 13th iteration. The changes in Kp and Ki are also shown in the table. The PI controller tuned with BA is given in (19).

$$G_{BEES} = 3.67 + 1601.4\frac{1}{s} \quad (19)$$

Also, two different conventional methods were used to determine the coefficients of the PI controller. The first of these is the Tyreus-Luyben method. Tyreus

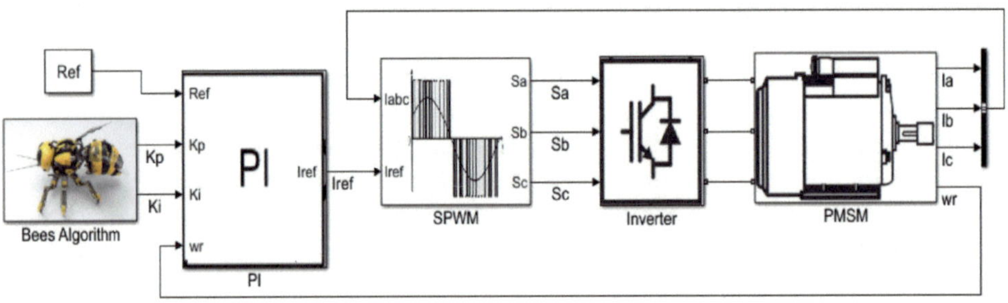

Figure 8.
Model of PMSM control with PI&SPWM.

It	Kp	Ki	Cost
1	1830,1931	9612,1901	8,47
2	1830,1931	9612,1901	8,47
3	1830,1931	9612,1901	8,47
4	63,7641	9199,0136	8,3991
5	63,7641	9199,0136	8,3991
6	17,6877	9119,7832	8,1792
7	17,6877	9119,7832	8,1792
8	17,6877	9119,7832	8,1792
9	16,6152	8932,0877	8,1632
10	16,6152	8932,0877	8,1632
11	16,6152	8932,0877	8,1632
12	7,59767	8919,6934	8,0425
13	3,6669	1601,3678	7,9262
14	3,6669	1601,3678	7,9262
15	3,6669	1601,3678	7,9262
16	3,6669	1601,3678	7,9262
17	3,6669	1601,3678	7,9262
18	3,6669	1601,3678	7,9262
19	3,6669	1601,3678	7,9262
20	3,6669	1601,3678	7,9262

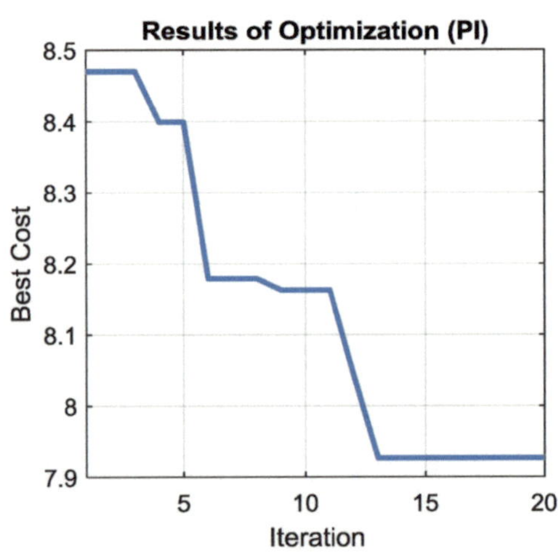

Figure 9.
Results of optimization (PI).

and Luyben's adjustment method is based on oscillations as in the Ziegler-Nichols method but has been modified for the controller parameters to achieve better stability in the control loop compared to the Ziegler-Nichols method. First, only P control is used and all gains are set to zero. The proportional gain K_p is increased until there are oscillations in the system response. K_p is increased until the oscillations are symmetrical. The final K_p value is recorded as Ku. The oscillation period of the signal is taken as Pu. K_p and T_i values are found according to the equation given in (20) [34].

$$K_p = 0.31K_u, T_i = 2.2T_u \tag{20}$$

When the K_p value is 27, the oscillations approached the symmetrical state (**Figure 10**). K_p and T_i values are calculated using (20).

$$\mathrm{K_p} = 0.31 * 27 = \mathbf{8.37}, \mathrm{T_i} = 2.2 * (0.01284 - 0.0122) = \mathbf{0.0014}$$

$$G_{TL} = Kp\left(1 + \frac{1}{T_{is}}\right) = 8.37 + 5944\frac{1}{s} \tag{21}$$

As the second conventional method, The Good Gain (TGG) method was used, which is more stable than the ZN method and therefore can obtain fewer oscillation values. In the TGG method, K_i value is chosen as 0 first and K_p value is increased starting from 0. This increase is continued until the answers close to the desired reference value are obtained. As shown in the graph below, when the peak value of the system response approaches the reference value, the K_p increase is stopped and the T_{ou} value is calculated. T_{ou} value is the time between the overshoot and undershoot values in the system response. From the T_{ou} value, the T_i and K_p values are calculated as shown below [35].

$$T_i = 1.5T_{ou}, K_p = 0.8K_p \tag{22}$$

When the K_p value is 4, the system response approaches the reference value (**Figure 10**). K_p and T_i values are calculated using (22).

$$\mathrm{K_p} = 4 * 0.8 = \mathbf{3.2}, \mathrm{T_i} = 1.5 * (0.00156 - 0.00116) = \mathbf{0.0006}$$

$$G_{TGG} = Kp\left(1 + \frac{1}{T_{is}}\right) = 3.2 + 5333\frac{1}{s} \tag{23}$$

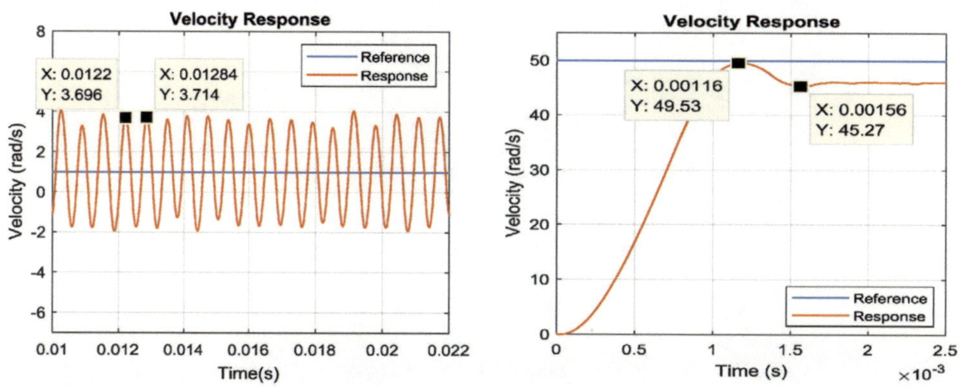

Figure 10.
Velocity response for Tyreus-Luyben method and the good gain method.

5. Experimental studies and results

To drive the PMSM, a motor driver card with a dsPIC33f model MCU is used. In addition to the MCU, the driver board includes the communication interface, the MOSFET H Bridge and drivers, and sensor reading interfaces. The driver board and other hardware used in the tests can be seen in **Figure 11**. Target Language Compiler in Simulink was used to convert the model into a machine code that dsPIC MCU can run directly. Firstly, ANSI C code was created and then machine code was generated by using the C30 C compiler provided by Microchip. Current sensors are located on the driver circuit in series with the motor phases. It is collected from each phase separately. Position data are taken from the digital encoder connected to the back of the motor shaft.

In experimental studies firstly, "100 rad/s" step command was applied to the PMSM for all controllers. All velocity results are given in **Figure 12**. and current results are given in **Figure 13**. It is seen that all four controllers are able to give enough response to the velocity command. But as can be seen, there is an overshoot (MPO) of 7% and 11.8% in PI controllers which were tuned by conventional methods. On the other hand, an overshoot of 4.2% was observed in PI tuned with BA. There is no overshoot in MPC. Because of the prediction realized by MPC using the PMSM model, controller output is produced in a more controlled manner after each step, thus creating a smooth effect on the system. In PI controllers, rise time (RT) is 0.1 ms smaller than MPC. The aggressiveness seen in overshoot is also seen here. On the other hand, in the settling time (ST) smaller value was obtained in MPC. The most important factor here is the overshoot and oscillations in PI controller responses. The resistance and inductance values of PMSM used in this study are very low. For this reason, high currents can be seen especially during take-off. Because of the optimum switching with MPC, instantaneous accelerations and currents can be suppressed. The low-value fluctuations seen in the velocity responses in PI control, after the settling time, cause continuous current to be drawn from the battery. **Table 1** shows the performance values of the controllers.

As the second test, a sinusoidal signal with an amplitude of 100 rad/s and a frequency of 2 Hz was applied to the PMSM. Test results are shown in **Figures 14** and **15**. MPC also follows the reference value in this test with a steady state error of

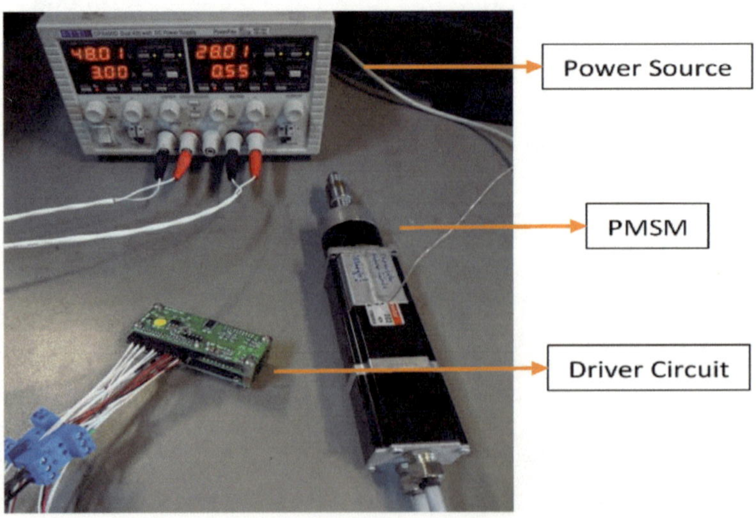

Figure 11.
Experimental environment.

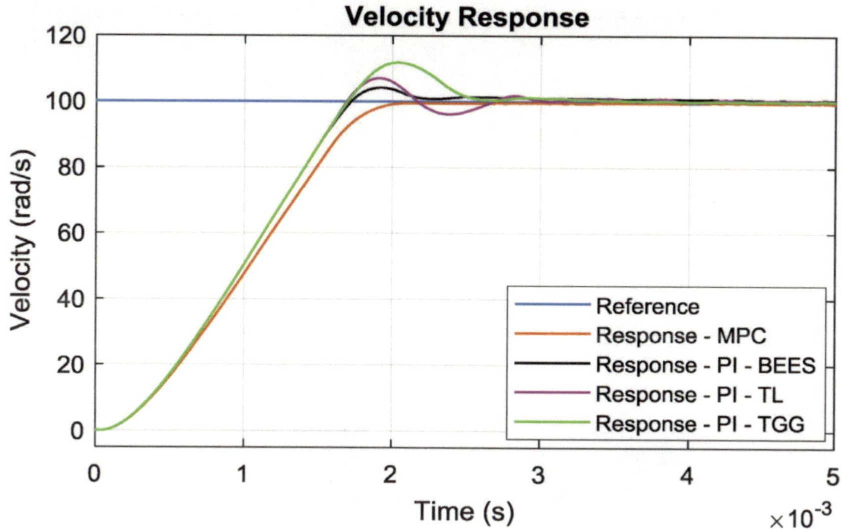

Figure 12.
Velocity responses of controllers (step response).

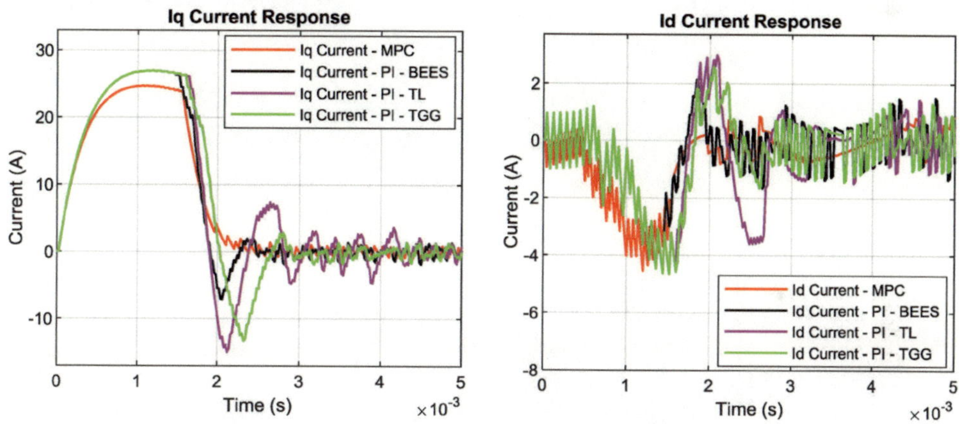

Figure 13.
Iq & id Current Responses of controllers (step response).

Controller	Results of step response				
	SS error	RT	ST	MPO	Max. Iq current
MPC	0.3%	1.3 ms	1.86 ms	0%	24.7 A
PI - BEES	0.5%	1.2 ms	2.2 ms	4.2%	26.9 A
PI - TGG	0.1%	1.2 ms	2.42 ms	11.8%	26.9 A
PI - TL	0.3%	1.2 ms	2.52 ms	7%	26.9 A

Table 1.
Results of step response.

approximately 0.4%. In the MPC cost function design, especially low power consumption was emphasized and power constraints were added. Similarly, in the calculation of the weights with BA, the bus current is also used with the speed error in MOF. Thus, when calculating the weights, the values with low speed error and

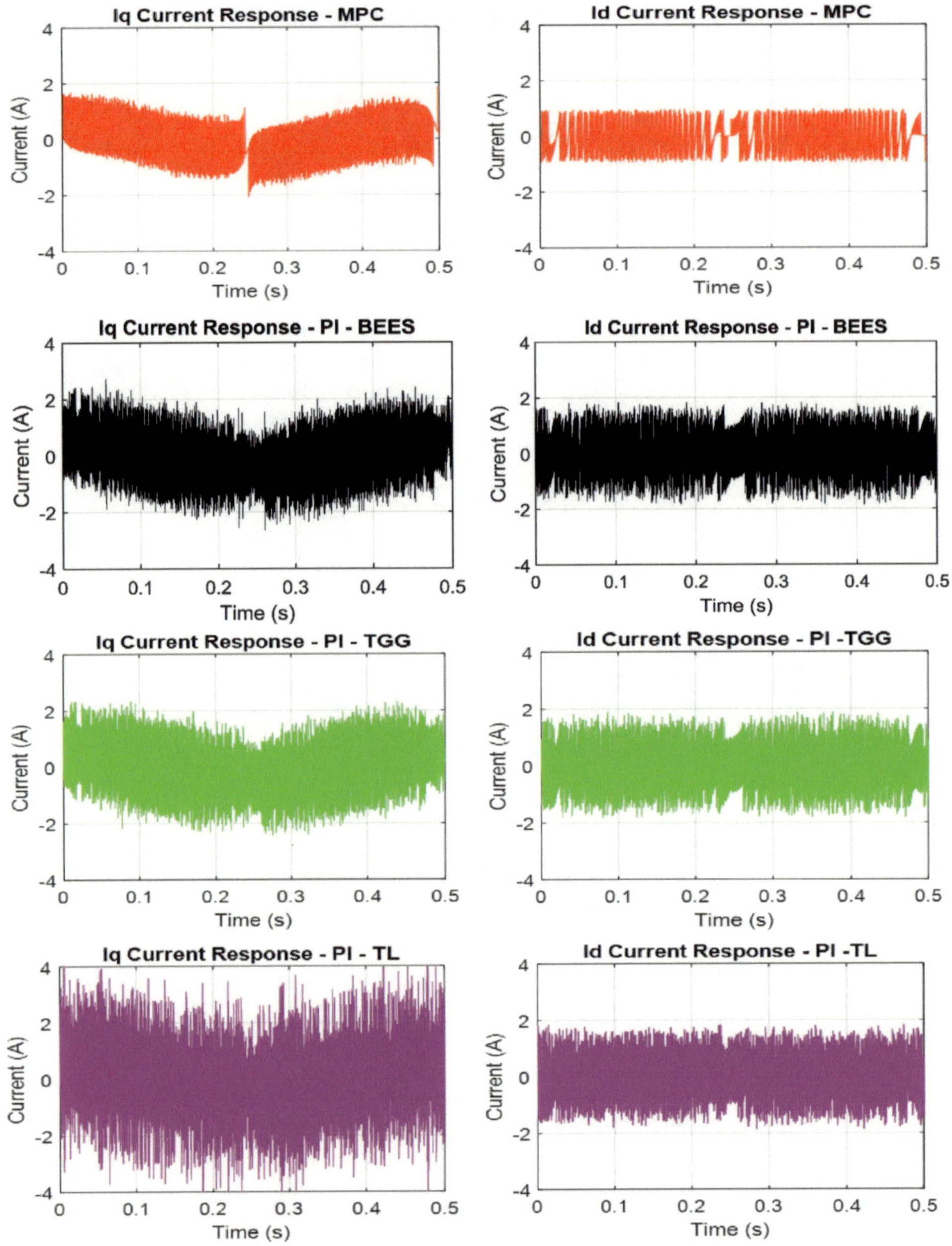

Figure 14.
Iq & id Current Responses of controllers.

low currents at the same time came to the fore. Therefore, a very low steady state error occurs in MPC control.

In PI controllers, although there is no steady state error, higher oscillations are seen compared to MPC. These oscillations can cause significant damage to systems, especially in applications requiring precise control. The effect of oscillations in the velocity response is clearly seen in the flow results. Amplitudes less than 2 A in MPC is up to 4 A in PI controllers.

Finally, a position control application was implemented to test the controllers. Only a P controller with a gain value of 100 has been added to existing controllers. As a test signal, a profile with many changes of direction was used to test its

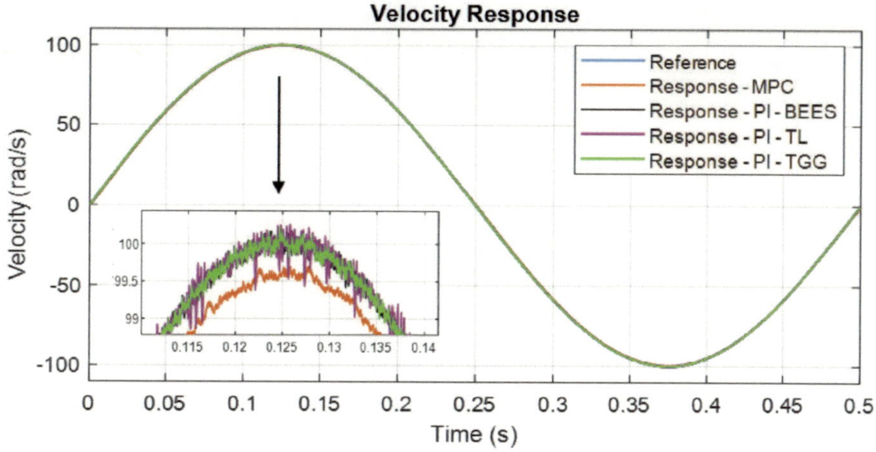

Figure 15.
Velocity responses of controllers (sinus response).

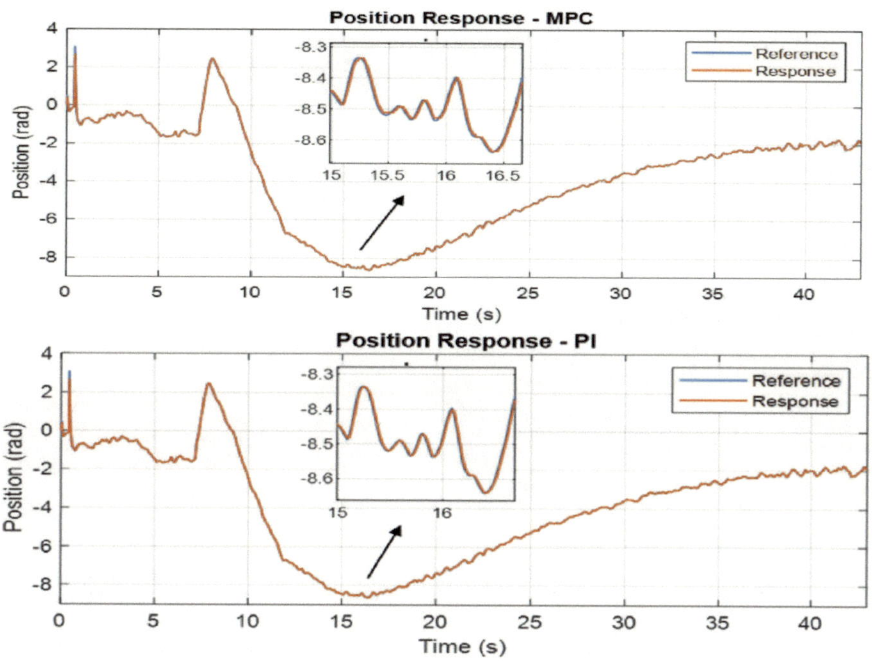

Figure 16.
Position responses of controllers.

performance with the switch for the PI coefficients, those tuned with BA were used. Position results are shown in **Figure 16** and position errors are shown in **Figure 17**.

Both controllers were able to respond to the references successfully. Error values are around 0.02 rad, except for the instantaneous step reference applied initially. Current data can also be seen in **Figure 18**. As with velocity application, MPC control uses less current in position control. It is seen that the optimum inverter switching method used in this study gives successful results. The success of the weights calculated with BA in PMSM control has also been confirmed by the position control application.

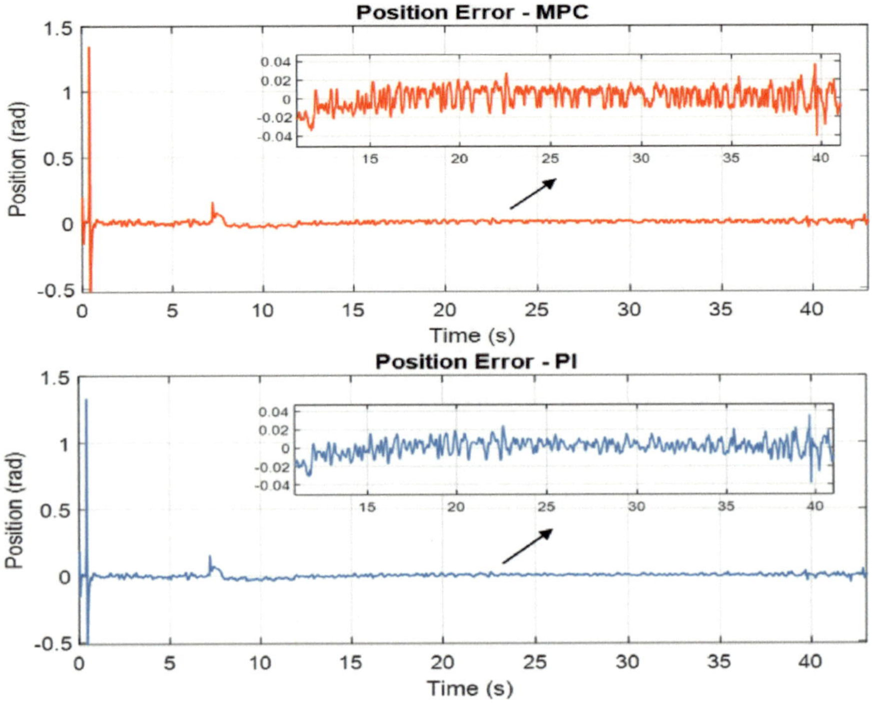

Figure 17.
Position errors of controllers.

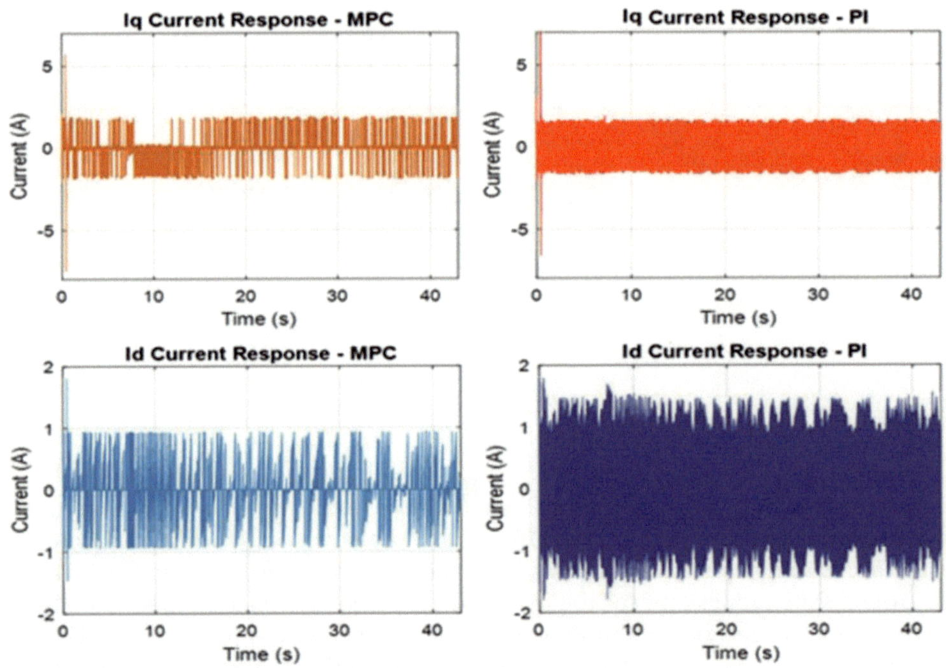

Figure 18.
Current responses of position controllers.

6. Conclusion

FCS-MPC has a limited number of optimization and calculation processes. Therefore, the delay compensation method was used to prevent timing errors in MPC. Thus, in this application, MPC was used with a relatively low-level processor without any problem. In addition to speed error and current predictions, power prediction has been added to the standard cost function used for speed control. In this way, in each switching selection, the possibilities that the lowest power consumption may occur along with other factors are evaluated.

For the Bees Algorithm, which is used to determine the weight coefficients, an infrastructure has been established to minimize the speed error and bus current. With the fast search capability of BA, optimum weight coefficients were calculated in approximately 15 iterations. Because of the low current and low energy preferences used in both the MPC and BA, MPC has achieved a more effective and less oscillatory control by using much lower currents than PI methods. With MPC, PMSM has been controlled with 15% settling time than other controllers and also with no overshoot. There is no exact method for determining the weight coefficients. It seems that manual adjustment is still preferred in many applications. With the efficient neighborhood search structure of BA, weights can be calculated with a small number of iterations. It provides great convenience for researchers. By designing a multi objective function, the number of variables that can be optimized can be increased if desired.

Energy consumption in autonomous vehicles and robots is one of great importance. MPC is used in this context with its smooth control structure. As in this application, BA can be preferred for autonomous control applications that require weight optimization. One of the advantages of BA over other meta-heuristic algorithms is that there is no mathematical equation in its structure. In this way, it can be used easily on different platforms and software languages.

In future works, the parameters of the CARIMA method, which is one of the popular MPC methods, will be optimized using this algorithm. A comparison of different MPC methods will be carried out together with the same test system.

Acknowledgements

The author would like to thank Roketsan A.S. for their financial support for this work.

Conflict of interest

I declare that this manuscript is original, has not been published before, and is not currently being considered for publication elsewhere. I know of no conflicts of interest associated with this publication, and there has been no significant financial support for this work that could have influenced its outcome.

Appendix - A

Model Configuration
Solver type: Fixed step

Solver: ode 5 (Dormand–Prince)
Fixed-step size: 2e-5 [s]
Tasking mode: Singletasking
PMSM Parameters:
R = 0.894;% Resistance [Ohm]
Ts = 2e-5; % Sampling time [s]
L = 0.338e-3;% Inductance [H]
Fl = 0.0329; % Flux linkage [Wb]
Vdc = 48;% DC-link voltage [V]
J = 368e-7; % Inertia [kg.m2]
p = 2; % Pole pairs

Appendix - B

Bees Algorithm Parameters.
MaxIt = 20;% Maximum Number of Iterations
nScoutBee = 20;% Number of Scout Bees
nBestSite = 4;% Number of Best Sites
nEliteSite = 2;% Number of Elite Sites
nBestSiteBee = 5;% Number of Recruited Bees for Best Sites
nEliteSiteBee = 10;% Number of Recruited Bees for Elite Sites

Author details

Murat Sahin
Control Systems Department, Roketsan A.S., Ankara, Turkey

*Address all correspondence to: msahin@roketsan.com.tr

References

[1] Cai, R., Zheng, R., Liu, M. and Li, M. Robust Control of PMSM Using Geometric Model Reduction and u-Synthesis. IEEE Transactions on Industrial Electronics, vol. 65, no. 1, pp. 498-509, Jan. 2018, doi: 10.1109/TIE.2017.2714140..

[2] Wang, Z., Chen, J., Cheng, M., Chau, K. T. Field-Oriented Control and Direct Torque Control for Paralleled VSIs Fed PMSM Drives With Variable Switching Frequencies. IEEE Transactions on Power Electronics, Volume: 31, Issue: 3, 2016; pp. 2417 – 2428. Doi: 10.1109/TPEL.2015.2437893.

[3] Wang, Z., Yu, A., Li, X., Zhang, G., and Xia, C. A Novel Current Predictive Control Based on Fuzzy Algorithm for PMSM. IEEE Journal of Emerging and Selected Topics in Power Electronics, vol. 7, no. 2, pp. 990-1001, June 2019, doi: 10.1109/JESTPE.2019.2902634.

[4] Feng, G., Lai, C., Kar, N. C. A Closed-Loop Fuzzy-Logic-Based Current Controller for PMSM Torque Ripple Minimization Using the Magnitude of Speed Harmonic as the Feedback Control Signal. IEEE Transactions on Industrial Electronics Volume: 64, Issue: 4, 2017; pp. 2642 – 2653. Doi: 10.1109/TIE.2016.2631524.

[5] Mani, P., Rajan, R., Shanmugam, L., and Joo, Y. H. Adaptive Fractional Fuzzy Integral Sliding Mode Control for PMSM Model. IEEE Transactions on Fuzzy Systems, vol. 27, no. 8, pp. 1674-1686, Aug. 2019, doi: 10.1109/TFUZZ.2018.2886169.

[6] Slapak, V., Kyslan, K., Durovsky, F. Position Controller for PMSM Based on Finite Control Set Model Predictive Control. Elektronika Ir Elektrotechnika, ISSN 1392-1215, VOL. 22, NO. 6, 2016; pp. 17-21. Doi: 10.5755/j01.eie.22.6.17217.

[7] Šlapák, V., Kyslan, K., Lacko, M., Fedák, V., Durovský, F. Finite Control Set Model Predictive Speed Control of a DC Motor. Hindawi Publishing Corporation Mathematical Problems in Engineering, Volume 2016, Article ID 9571972, 2016; 10 pages. Doi: 10.1155/2016/9571972.

[8] Metry, M., Balog, R. S. A Parameter Mismatch Study on Model Predictive Control Based Sensorless Current Mode. IEEE Texas Power and Energy Conference (TPEC). 2018, Doi: 10.1109/TPEC.2018.8312065.

[9] Espinoza, J., Buele, J., Castellanos, E. X., Pilatásig, M., Ayala, P., García, M. V. Real-Time Implementation of Model Predictive Control in a Low-Cost Embedded Device. Systemics, Cybernetics and Informatics Volume 16 - Number 2, 2018; pp. 72-77.

[10] Bartsch, A. G., Negri, G. H., Cavalca, M. S. M., Oliveira, J., and Nied, A. Cost function tuning methodology for FCS-MPC applied to PMSM drives. 2017 Brazilian Power Electronics Conference (COBEP), 2017, pp. 1-6, doi: 10.1109/COBEP.2017.8257320.

[11] Mahmoudi, H., Aleenejad, M., Moamaei, P., Ahmadi, R. Fuzzy Adjustment of Weighting Factor in Model Predictive Control of Permanent Magnet Synchronous Machines Using Current Membership Functions. IEEE Power and Energy Conference at Illinois (PECI), 2016; Doi: 10.1109/PECI.2016.7459225.

[12] Rodríguez, J., Kennel, R. M., Espinoza, J. R., Trincado, M., Silva, C. A., Rojas, C. A. High-Performance Control Strategies for Electrical Drives: An Experimental Assessment. IEEE Transactions on Industrial Electronics, Vol. 59, No. 2, 2012; pp. 812-820. Doi: 10.1109/TIE.2011.2158778.

[13] Codreş, B., Găiceanu, M., Şolea, R., Eni, C. Model Predictive Speed Control of Permanent Magnet Synchronous Motor. International Conference on Optimization of Electrical and Electronic Equipment (OPTIM), 2014; pp. 477-482. Doi: 10.1109/OPTIM.2014.6850946.

[14] Dani, S., Sonawane, D., Ingole, D., Patil, S. Performance Evaluation of PID, LQR and MPC for DC Motor Speed Control. 2nd International Conference for Convergence in Technology (I2CT), 2017; pp. 348-354. Doi: 10.1109/I2CT.2017.8226149

[15] Ma, H., Chu, L., Guo, J., Wang, J., and Guo, C. Cooperative Adaptive Cruise Control Strategy Optimization for Electric Vehicles Based on SA-PSO With Model Predictive Control. IEEE Access, vol. 8, pp. 225745-225756, 2020, doi: 10.1109/ACCESS.2020.3043370.

[16] Zanchetta, P. Heuristic Multi-Objective Optimization for Cost Function Weights Selection in Finite States Model Predictive Control. Workshop on Predictive Control of Electrical Drives and Power Electronics, 2011; pp. 70-75. Doi: 10.1109/PRECEDE.2011.6078690.

[17] Mohammadi, A., Asadi, H., Mohamed, S., Nelson, K., Nahavandi, S. Multiobjective and Interactive Genetic Algorithms for Weight Tuning of a Model Predictive Control-Based Motion Cueing Algorithm. IEEE Transactions on Cybernetics, Vol. 49, No. 9, 2019; pp. 3471-3481. Doi: 10.1109/TCYB.2018.2845661.

[18] Zhou, Z.D., Xie, Y. Q., Pham, D. T., Kamsani, S., and Castellani, M. Bees Algorithm for multimodal function optimization. Proceedings of the Institution of Mechanical Engineers, Part C: Journal of Mechanical Engineering Science, 2016, Volume: 230 issue: 5, page(s): 867-884. doi.org/10.1177/0954406215576063. .

[19] Zarchi, M. & Attaran, B. Performance improvement of an active vibration absorber subsystem for an aircraft model using a bees algorithm based on multi-objective intelligent optimization. 2017, Engineering Optimization, 49:11, 1905-1921, DOI: 10.1080/0305215X.2017.1278757. .

[20] Moradia, A., Nafchib, A. M., Ghanbarzadeh, A. Multi-objective optimization of truss structures using the bee algorithm. Scientia Iranica B (2015) 22(5), 2015; pp. 1789-1800.

[21] Kizir, G., Demirbas, S., Sahin, M. Real-time resistance estimation of permanent magnet synchronous motor. 2018 IEEE 12th International Conference on Compatibility, Power Electronics and Power Engineering (CPE-POWERENG 2018); pp. 1-5. Doi: 10.1109/CPE.2018.8372532.

[22] Rodriguez, J., and Cortes, P. Predictive Control of Power Converters and Electrical Drives. John Wiley & Sons, Ltd., Publication, 2012; pp. 133-143. ISBN: 978-1-119-94264-1.

[23] Wang, F. Model Predictive Torque Control for Electrical Drive Systems with and without an Encoder. PhD Thesis, Technical University of Munich, 2014; pp. 7-10.

[24] Jin, T., Shen, X., Su, T., Flesch, R. C. C. Model Predictive Voltage Control Based on Finite Control Set With Computation Time Delay Compensation for PV Systems. IEEE Transactions on Energy Conversion (Volume: 34, Issue: 1, March 2019); pp. 330-338. Doi: 10.1109/TEC.2018.2876619.

[25] Khosravi, M., Amirbande, M., Khaburi, D. A., Rivera, M., Riveros, J., Rodriguez, J., Vahedi, A., and Wheeler, P. Review of model predictive control strategies for matrix converters. IET Power Electronics, 2019, Volume 12, Issue 12 p. 3021-3032, doi.org/10.1049/iet-pel.2019.0212. .

[26] Baronti, L., Castellani M., Pham D. T. An analysis of the search mechanisms of the bees algorithm. Swarm and Evolutionary Computation 59, 2020; 100746. Doi: 10.1016/j.swevo.2020.100746.

[27] Liu, J., Zhou, Z., Pham, D. T., Xu, W., Yan, J., Liu, A., Ji, C, Liu, Q. An improved multi-objective discrete bees algorithm for robotic disassembly line balancing problem in remanufacturing. The International Journal of Advanced Manufacturing Technology, 2018; 97: 3937–3962.

[28] Ismail, A. H., Hartono, N., Zeybek, S., Pham, D. T. Using the Bees Algorithm to solve combinatorial optimisation problems for TSPLIB. IOP Conf. Series: Materials Science and Engineering 847, 2020; 012027.

[29] Castellani, M, Otri, S., and Pham, D. T. Printed circuit board assembly time minimisation using a novel Bees Algorithm. Computers & Industrial Engineering, Volume 133, 2019, Pages 186-194, ISSN 0360-8352, https://doi.org/10.1016/j.cie.2019.05.015.

[30] Zheng, J. H., Kou, Y. N., Jing, Z. X., and Wu, Q. H. Towards Many-Objective Optimization: Objective Analysis, Multi-Objective Optimization and Decision-Making. IEEE Access, vol. 7, pp. 93742-93751, 2019, doi: 10.1109/ACCESS.2019.2926493..

[31] Gunantara, N. A review of multi-objective optimization: Methods and its applications. Journal Cogent Engineering, Volume 5, 2018 - Issue 1, 2018; pp. 1-16. Doi: 10.1080/23311916.2018.1502242.

[32] Hruska, K.; Dvorak, P. Optimization of a PMSM Design for Control with Zero Direct Axis Current Component. 2016 ELEKTRO, 10.1109/ELEKTRO.2016.7512057. 2016; Doi: 10.1109/ELEKTRO.2016.7512057.

[33] Namboodiri, A., Wani, H. S. Unipolar and Bipolar PWM Inverter. IJIRST –International Journal for Innovative Research in Science & Technology| Volume 1, Issue 7. 2014; ISSN (online): 2349-6010.

[34] Haugen, F. Comparing PI Tuning Methods in a Real Benchmark Temperature Control System. Modeling, Identification and Control, Vol. 31, No. 3, 2010; pp. 79-91, ISSN 1890-1328. Doi: 10.4173/mic.2010.3.1.

[35] Haugen, F. The Good Gain method for simple experimental tuning of PI controllers. Modeling, Identification and Control, Vol. 33, No. 4, 2012; pp. 141–152, ISSN 1890–1328. Doi: 10.4173/mic.2012.4.3.